Traffic Engineering Design

WITHDRAWN-UN

D1530579

WITHDRAWN-ULS

Traffic Engineering Design

Traffic Engineering Design

Principles and Practice

Mike Slinn[1], Peter Guest[2] and Paul Matthews[3]

[1,3]MVA Consultancy, UK and [2]Hill Cannon Partnership, UK

A member of the Hodder Headline Group
LONDON • SYDNEY • AUCKLAND

First published in Great Britain in 1998 by
Arnold, a member of the Hodder Headline Group,
338 Euston Road, London, NW1 3BH

http://www.arnoldpublishers.co.uk

© Mike Slinn, Peter Guest and Paul Matthews

All rights reserved. No part of this publication may be reproduced or transmitted in any
form or by any means, electronically or mechanically, including photocopying, recording
or any information storage or retrieval system, without either prior permission in writing
from the publisher or a licence permitting restricted copying. In the United Kingdom such
licences are issued by the Copyright Licensing Agency: 90 Tottenham Court Road,
London W1P 9HE.

British Library Cataloguing in Publication Data
A catalogue record for this book is available from the British Library

Library of Congress Cataloging-in-Publication Data
A catalog record for this book is available from the Library of Congress

ISBN: 0 340 67647 7

Publisher: Eliane Wigzell
Production Editor: Wendy Rooke
Production Controller: Priya Gohil
Cover designer: Stefan Brazzo

Typeset in 10/12pt Times by Phoenix Photosetting, Chatham, Kent
Printed and bound in Great Britain by JW Arrowsmith Ltd, Bristol

Contents

Acknowledgements

The authors wish to acknowledge the following for their assistance and permission to use materials:

  Department of Transport, Environment and the Regions (DETR)
  London Transport
  Hampshire County Council
  West Sussex County Council
  Birmingham City Council
  Leeds City Council
  London Borough of Southwark
  Luton Borough Council
  Institution of Highways and Transportation
  MVA Limited
  Hampshire Constabulary
  Dr J Miles
  Langdale Systems
  Metric Parking
  Parking Technology

1
Introduction

1.1 Original version

In 1970, the first version of this book, written by Gordon Wells, was published as an introduction to the then relatively new subject of traffic engineering. Since that date the book has been twice updated by the original author, who produced the last volume in 1978. However, in the last two decades the range of skills required by the modern traffic engineer has developed virtually beyond recognition and it became clear to the publishers that the book needed rewriting.

Gordon Wells has now moved on and is currently pursuing a successful career as an author of children's literature and so the task of attempting to bridge the gap has fallen to the present authors. We are all practitioners and have attempted, in producing this book, to meet the same objectives as the original when it was written some 30 years ago. The purpose of this volume is to provide the reader with a basic understanding of the range of skills and techniques needed by the modern traffic engineer.

This book is an introduction to the subject and as such cannot be exhaustive. Indeed, techniques are developing so rapidly in some areas that some parts of the book may be superseded by new techniques, particularly in the appliance of technology, almost before the book is published. However, as practitioners, we have sought to set out the basics in a clear and easy to understand style.

1.2 What is traffic?

Traffic is not a new phenomenon. Roman history records that the streets of Rome were clogged with traffic, and at least one emperor was forced to issue a proclamation threatening the death penalty to those whose chariots and carts blocked the way. More recently, pictures of our modern cities at the turn of the century show city streets clogged with traffic.

What do we mean by traffic? The dictionary describes 'traffic' as the transportation of goods, coming and going of persons or goods by road, rail, air etc. Often in common usage we forget this wider definition and colloquially equate the word with motorised road traffic, to the exclusion of pedestrians and even cyclists. Traffic engineering is concerned with the wider definition of traffic and this book deals with the design of facilities for all forms of traffic on surface modes of transport. This excludes aviation and shipping.

1.3   What is traffic engineering?

In the introduction to his book Gordon Wells quoted the Institution of Civil Engineers[1] for his definition of traffic engineering; this is:

> That part of engineering which deals with traffic planning and design of roads, of frontage development and of parking facilities and with the control of traffic to provide safe, convenient and economic movement of vehicles and pedestrians.

Although this definition is still a valid one, there has clearly been a change in the emphasis in the role of the traffic engineer in the last decade. This change has been in response to changes in both society's expectations and concerns about traffic and the impact of traffic on the wider environment. There has also been a pragmatic change forced on traffic engineers as traffic growth has continued, unabated, and so the engineer has been forced to fit more traffic onto a finite highways system.

Since 1970, road travel in the UK has increased by about 75% and, although many new roads have been built, these have tended to be interurban or bypass roads, rather than new roads in urban areas. Thus, particularly in urban areas, the traffic engineer's role is, increasingly, to improve the efficiency of an existing system rather than to build new higher capacity roads.

1.4   How much traffic?

By the end of the first world war there were about one-third of a million motor vehicles in the UK. Within 6 years this number had increased by a factor of nearly five to 1.5 million vehicles. Table 1.1 shows the growth in traffic since then.

The rapid increase in vehicle ownership (by 42%) in the last 20 years is clear. Table 1.2 compares growth in the UK with that in other countries. However, the change in the numbers of vehicles does not tell the whole story.

Not only do more people own cars; each vehicle is used more. In addition, the pattern of freight movement has changed dramatically, with a shift from rail to road and radical changes in distribution procedures, which mean that goods now tend to be distributed from fewer, larger depots, with a consequent increase in goods vehicle travel. Table 1.3 shows the increase in travel on the roads system since 1975.

During the same period, the highway network was increased by the construction of new roads, such as bypasses and, latterly by motorways. Table 1.4 shows the increase in network size, between 1987 and 1995.

Table 1.1 Growth in UK vehicles[2]

Year	Number of vehicles
1919	330 000
1925	1 510 000
1935	2 611 000
1945	2 606 000
1955	6 624 000
1965	13 259 000
1975	17 884 000
1985	21 159 000
1995	25 369 000

In the UK, the Department of Transport uses a method of predicting future traffic which links growth in car ownership to, amongst other things, predicted changes in economic performance.[3,4] The methodology provides high and low predictions based on expectations about economic performance. The predictions, which have, if anything, proved to be historically conservative, suggest that by the year 2020 traffic levels will more than double.

Table 1.2 Domestic versus overseas traffic growth[a]

	Cars and taxis		Goods vehicles[1]		Motor cycles etc.[2]		Buses and coaches		Total	
	1984	1994	1984	1994	1984	1994	1984	1994	1984	1994
Great Britain	16 775	21 231	1 769	2 438	1 419	757	149	154	20 112	24 580
Northern Ireland	439	509	37	59	16	9	2	5	494	582
United Kingdom	17 214	21 740	1 806	2 497	1 435	766	151	158	20 606	25 161
Belgium	3 300	4 210	232	403	498	187	12	15	4 042	4 815
Denmark	1 440	1 610	239	313	200	49	8	13	1 887	1 985
France	20 800	24 900	2 868	3 606	5 065	2 561	62	79	28 795	31 146
Germany	28 375	39 765	1 497	2 114	4 223	3 750	124	71	34 219	45 700
Greece	1 151	1 959	572	808	154	388	18	23	1 895	3 178
Irish Republic	717	939	84	136	26	24	3	5	830	1 104
Italy	20 888	30 420	1 683	2 543	5 163	5 397	72	78	27 806	38 438
Luxembourg	146	229	9	13	3	8	1	1	159	271
Netherlands	4 841	5 884	345	565	785	844	12	12	5 983	7 305
Portugal	1 265	3 532	347	442	102	192	10	11	1 724	4 178
Spain	8 874	13 441	1 643	2 832	706	1 279	41	47	11 264	17 599
Austria	2 468	3 479	203	283	646	742	9	10	3 326	4 514
Croatia	–	698	–	42	–	9	–	9	–	759
Czech Republic	2 640	2 967	260	161	612	476	35	23	3 547	3 627
Finland	1 474	1 873	171	249	211	158	9	8	1 865	2 288
Hungary	1 344	2 179	142	245	394	157	25	22	1 905	2 602
Norway	1 430	1 633	90	319	176	161	16	29	1 712	2 142
Slovak Republic	–	994	–	148	–	229	–	12	–	1 383
Sweden	3 081	3 594	210	304	22	57	14	14	3 327	3 969
Switzerland	2 552	3 165	193	250	847	708	12	14	3 604	4 137
Japan	27 144	40 772	17 010	22 246	17 354	16 396	230	248	61 738	79 662
USA	127 867	133 930	38 047	63 445	5 480	3 718	584	670	171 978	201 763

[a] Numbers represent thousands.

Table 1.3 Change in road traffic in the UK[2]

Year	Estimated number of vehicle km $\times 10^9$
1975	244.4
1985	309.7
1995	430.9

Table 1.4 Change in the highway network in the UK[2]

Year	Network length (km)
1985	348 699
1995	366 999

1.5   **Structure of this book**

This book has been prepared in the context of traffic engineering as practised in Britain, with references to standard UK designs and legislation. That said, many of the basic principles are the same, regardless of the country in which they are applied. The authors hope, therefore, that the information contained in the book will be of general interest to a wider audience who will be able to use and adapt the information contained in this book to their circumstances.

The book is set out to follow a logical sequence of steps designed to allow the reader to first measure and understand 'traffic' and then to design measures to deal with and control it.

Chapters 2 and 3 deal with surveys. Before attempting to undertake any task, it is important to obtain a measure of, at least, the base level of traffic, to ensure that any solution is appropriate and of correct scale. Chapter 2 describes a range of survey methods for measuring traffic flow on a road system whereas Chapter 3 deals with parking surveys.

Sometimes a solution can only be implemented in response to existing conditions. For example, a car park control scheme may be appropriate for today's conditions but may become obsolete as circumstances change. It is not always possible to predict the future and so, for some problems, it is only appropriate to design a scheme which deals with the existing situation.

Therefore, an understanding of existing traffic conditions is fundamental. Often, however, it is equally important to ensure that any traffic engineering solution will be capable of dealing with both existing traffic and the traffic expected throughout the design life of the project. For example, a junction design should take account of both present-day flow and expected growth for an agreed period. Indeed, the basis of traffic prediction of future traffic, from a new development is a fundamental part of the transport impact analysis process. The techniques for predicting traffic flow are dealt with in Chapter 4.

Chapter 5 deals with the concepts of traffic capacity and provides definitions of capacity and flow. The chapter describes techniques for estimating the effects of traffic flow on junction and highway performance.

This is followed by a number of chapters which deal with the fundamentals of traffic management and control, covering issues such as the design of roads and junctions, and signs and traffic signals. Chapter 9 deals with vehicle parking, both on- and off-street.

Road safety is a very important issue and a traffic engineer has a duty to ensure that any work he undertakes will result in a safe environment for all road users. Chapter 10 deals with road safety engineering and discusses concepts of designing in safety and measures to deal with pre-existing problems. This includes a description of the safety audit procedure which should now be a mandatory part of any new scheme.

Chapter 11 deals with traffic calming, describing how traffic engineering measures can be used to manage and reduce the adverse impacts of vehicular traffic, so as to provide a safer and better environment in existing streets. Chapter 12 deals with priority measures for public transport, including measures for the integration of light rail systems into road traffic.

Any new land use development, be it a housing estate, a new factory or an edge of town shopping centre, will generate traffic as people travel to and from the site. Chapter 13 discusses the current government view on development and sustainable transport and the techniques used to estimate the amount of traffic at different types of development, both during construction and when the development is complete. The chapter also describes how to measure both the community and environmental impacts that arise.

Chapter 14 develops the concept of sustainable transport further, dealing with the development control process and discusses measures designed to help promote cycling and walking and to minimise total travel demand.

The word 'telematics' has gained common usage in recent years, often used as a talisman to describe anything technical which has an application to traffic and travel. Unfortunately, many of the ideas which are so enthusiastically promoted by their developers suffer from inadequate development or are, frankly, solutions in search of a problem. Chapter 15 gives a sound basic grounding of how new technology is being applied successfully to improve the efficiency and safety of the transportation system, including a review of ideas that are likely to gain widespread use in the next few years.

Finally, Chapter 16 gives a brief overview of the current UK legislative framework under which the traffic engineer operates.

References

1. Wells, GR (1976) *Traffic Engineering An Introduction*, Charles Griffin & Company.
2. Department of Transport (1996) *Transport Statistics Great Britain*, HMSO, London.
3. Department of Transport (1996) *National Road Traffic Forecasts, 1989, Rebased at 1994*, HMSO, London.
4. Department of Transport (1989) Highways Economic Note 2, HMSO, London.

2
Traffic Surveys

2.1 Introduction

Traffic engineering is used to either improve an existing situation or, in the case of a new facility, to ensure that the facility is correctly and safely designed and adequate for the demands that will be placed on it.

In an existing situation we have to know the present day demands and patterns of movement, so that the new measure can be designed adequately. With a new road or facility, there is obviously no existing demand to base the design on; therefore, we have to estimate the expected demand.

If a new facility replaces or relieves existing roads, for example a bypass or a new cycle track, we can estimate the proportion of traffic that could be expected to transfer using a traffic assignment (see Chapter 4).

If the facility is completely new, for example a road in a new development, then the expected traffic and hence the scale of construction needed has to be estimated another way. This is usually done by a transport impact analysis (see Chapter 13) which will seek to assess the likely level of traffic by reference to the traffic generated by similar developments elsewhere. In either case the starting point will be a traffic survey.

The main reason for undertaking a traffic survey is to provide an objective measure of an existing situation. A survey will provide a measure of conditions at the time that the survey was undertaken. A survey does not give a definitive description of a situation for ever and a day and, if the results are to be used as representative of 'normal' traffic conditions, the survey must be defined with care and the information used with caution.

Traffic flow varies by time of day, day of the week and month of the year. Figure 2.1 shows a typical 24 hour daily flow profile for an urban area. The figure shows morning and evening peaks as people travel to and from work and flow drops off at night, to a lower level than observed either during the day, when commercial activity takes place, or in the evening, when social activities tend to take place.

Traffic flows also tend to vary by day of the week (Figure 2.2). Again, on a typical urban road traffic flows tend to build during the week, to a peak on Friday. Flows are lower at the weekend, when fewer people work and lowest on Sunday; although the introduction of Sunday trading has affected the balance of travelling at the weekend.

The variation in pattern of travel over the year depends a great deal upon location. In urban areas, which are employment centres, flow drops during the summer period when schools are

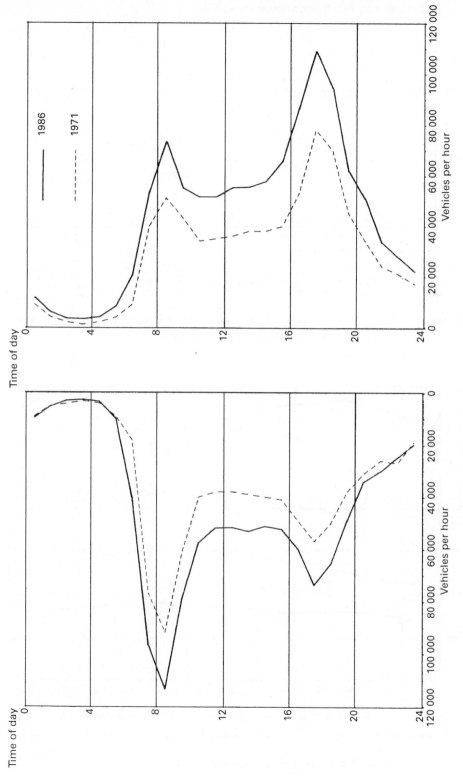

Fig. 2.1 Graph showing 24 hour flow profile.

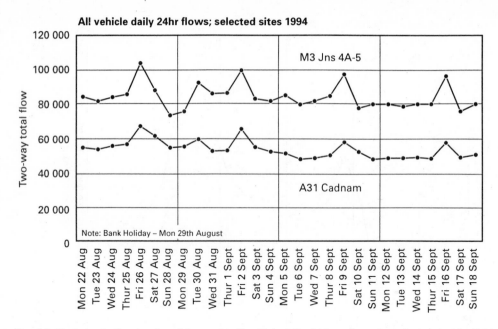

Fig. 2.2 Variation in flow by day of the week: all vehicle daily 24 hour flows for selected sites in 1994. Note that 29 August was a bank holiday.

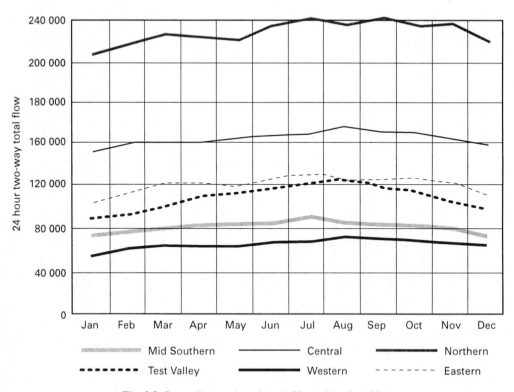

Fig. 2.3 Screenlines and cordons in Hampshire for 1994.

closed and workers tend to take annual holidays. This is balanced by a reverse trend in holiday areas, where traffic flows increase dramatically in July and August, and roads which are adequate most of the year become heavily congested. The effect can be less dramatic on interurban roads, other than those providing access to holiday areas, as, to an extent, the decline in interurban business travel during the summer is offset by tourism. Figure 2.3 shows examples of typical annual flow profiles.

The information above shows that the pattern of flow on any road can be highly variable and, in deciding when and where to undertake a traffic survey, it is important to ensure that the survey provides a fair measure of the traffic conditions that are being studied. To take the example of the road in a tourist area, a traffic survey on August bank holiday would measure peak traffic conditions. As these levels occur only one or two days a year there would be little point in using this data as a basis for design, as the scheme would be overdesigned for traffic conditions most of the time.

Generally, traffic surveys should not be planned to measure the peak of the peak but to measure the 'normal' peak conditions. Trunk road surveys may require a full year's survey of traffic so that the 50th (30th or 200th) highest hourly flow can be determined, and used as the basis for design.

2.2   How to define a traffic survey

The starting point in defining a traffic survey is to decide what question has to be answered and choose the type of survey accordingly. If the survey is not adequately planned, there is a danger that the wrong data will be collected and the traffic situation will not be correctly understood.

The only exception to this rule occurs when one is faced with a complex situation where it may not be possible, at first, to adequately understand what is going on, in terms of traffic flow and circulation. In these circumstances the traffic survey is providing evidence which will not only be used to quantify behaviour, it may also be used to define it.

2.3   Traffic counts

The traffic engineer has an increasing number of survey methodologies available to help him to understand traffic movement. The main techniques are described below, with their principal applications. All the traffic count methodologies described are noninterventionalist, that is they do not affect the traffic flow being measured.

2.3.1   AUTOMATIC TRAFFIC COUNTS

Automatic traffic counters are used to mechanically measure traffic volumes moving past the survey point. The counters normally use a pressure tube or an inductive loop which is fixed across the road at the census point. A pressure tube is compressed each time a vehicle axle crosses it. This sends a pulse along the tube which is counted and hence the vehicular flow can be estimated. More modern systems use a piezoelectronic tube and the electrical pulses are counted. Figure 2.4 shows a typical output from an automatic traffic counter with the data presented as hourly flow.

Week Beginning:- Thursday 24 April 1997
Site Reference:- CROY1
Vehicle Flow:- CHANNEL 2 - Outbound

HOUR ENDS	Thursday	Friday	Saturday	DAYS Sunday	Monday	Tuesday	Wednesday	5 DAY AVERAGE. VALUE %		7 DAY AVERAGE. VALUE %	
1	4	4	18	17	5	0	1	3	0.83	7	1.89
2	0	1	3	4	0	3	0	1	0.24	2	0.43
3	3	0	0	3	0	0	0	1	0.18	1	0.23
4	0	0	1	1	1	0	1	0	0.12	1	0.15
5	0	0	0	0	1	2	0	1	0.18	0	0.12
6	2	0	2	0	0	0	0	0	0.12	1	0.15
7	2	3	1	0	4	5	5	4	1.12	3	0.77
8	8	6	2	3	24	18	14	14	4.13	11	2.90
9	37	20	5	25	11	32	24	25	7.32	22	5.95
10	11	7	8	39	11	13	6	10	2.83	14	3.67
11	7	6	38	49	7	9	9	8	2.24	18	4.83
12	13	12	10	33	5	10	8	10	2.83	13	3.52
13	8	14	25	34	9	16	8	11	3.24	16	4.41
14	11	11	15	44	10	15	14	12	3.60	17	4.64
15	11	10	10	44	6	11	5	9	2.54	14	3.75
16	32	16	9	46	12	24	11	19	5.60	21	5.80
17	23	11	18	36	10	11	7	12	3.66	17	4.48
18	34	12	21	47	11	33	17	21	6.31	25	6.76
19	76	15	39	26	39	51	35	43	12.74	40	10.86
20	69	17	29	25	36	59	35	43	12.74	39	10.44
21	50	22	81	15	24	65	12	35	10.21	38	10.40
22	14	59	18	4	19	37	25	31	9.09	25	6.80
23	44	10	16	4	3	15	24	19	5.66	17	4.48
24	6	21	14	10	1	6	8	8	2.48	9	2.55

TOTALS

	Thursday	Friday	Saturday	Sunday	Monday	Tuesday	Wednesday	5 DAY	7 DAY
7 - 19	271	140	200	426	155	243	158	193	228
6 - 22	406	241	329	470	238	409	235	306	333
6 - 24	456	272	359	484	242	430	267	333	359
1 - 24	465	277	383	509	249	435	269	339	370

Fig. 2.4 Example of automatic traffic counter output.

A tube counter measures the impact of an axle and so traffic flow is derived from counting the number of impulses and dividing them by a factor representing the most common number of axles on a vehicle, i.e. two. On heavily trafficked roads where there area large number of multi-axle heavy vehicles, a slightly higher factor may be used. Inaccuracies can occur when two vehicles cross the loop at the same time, for example a motor cycle and a car, or when there is a higher than expected proportion of multi-axle vehicles. In high speed conditions, axle bounce, because of the road surface conditions, can cause failure to compress the tube.

An alternative is to use an inductive loop which will detect the mass of a vehicle. The passage of the metal mass of a vehicle over the loop induces a magnetic field in the loop, allowing the presence of a vehicle to be registered. This type of technology counts the vehicles' presence directly, with one pulse for each vehicle. The loops can give false readings if two vehicles pass

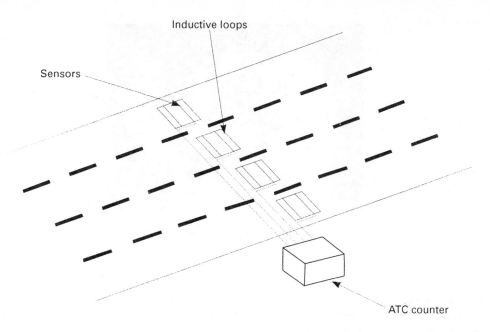

Fig. 2.5 Typical automatic traffic counter installation.

the loop simultaneously or are close together; conversely, a vehicle pulling a trailer can be read as two vehicles.

Automatic counters can also be set up to classify the type of vehicle by numbers of axles.

Automatic traffic counters are usually used where traffic flow data is required over an extended period, for example a week or a year. The data can be presented in terms of the flow per time period, for example, per hour, per day or per week, and used to compare daily, weekly or seasonal variation, as well as quantifying the volume of traffic. Automatic counters are useful when one wishes to collect indicative data over an extended period cheaply. The counters cannot give precise information.

For a counter which has been installed at a particular location for a short period of time, the data can be collected at the side of the road and stored on a data tape which can be collected periodically (Figure 2.5). If the counter is part of a permanent installation the data can be collected remotely using telemetry.

2.3.2   MANUAL COUNTS

Traffic flows can be measured by manual observation, instead of using an automatic counter. Traffic flowing past a survey point is counted by an observer, who records the flow using either a tally counter (Figure 2.6) or by taking a manual count of vehicles and recording it on paper, typically using a five-bar-gate counting technique or by using a hand-held computer.

Counts can be classified, to identify the volume and mix of types of vehicles using the road at the survey point. Figure 2.7, from the DOT publication *Roads and Traffic in Urban Areas*[1] shows a typical classification. However, the level of classification used will very much depend upon the

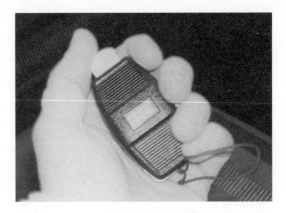

Fig. 2.6 Tally counter.

needs of the survey. For example, it may be adequate to use a simpler form of classification, such as cars and taxis, buses and commercial vehicles. The engineer should choose an appropriate level of classification for each study.

If a data collection survey is only planned to cover a short period of time, then the expense of installing an automatic counter may not be justified when compared with the cost of using a surveyor. The surveyor also has the ability to discriminate between classes of vehicles.

Manual counts generally offer better value for money when data is required for a single day, or for less than the full 24 hour day but collected over two or three days. Manual classified counts become more difficult where flows are very high, and where a break in concentration can introduce high error rates in the count. Figure 2.8 shows a typical survey instrument for a manual classified count.

If the engineer wishes to gain a quick insight to traffic conditions over a wider area, short period, sample traffic counts can be taken over a wide area and factored up, to represent the hourly flow. Thus, if one wished to have an understanding of traffic levels at a complex junction, traffic could be counted at each arm for 5–10 minutes and then factored up to hourly counts, to give an understanding of conditions. This is a good method of gaining a quick insight into traffic levels, but should not be used as a substitute for a properly organised traffic survey.

2.3.3   TURNING MOVEMENTS

A manual classified count (MCC) records directional traffic flow past a survey point. The survey point could be midlink or at a junction. If we wish to understand how traffic is behaving at a junction more precisely, we extend the complexity of the MCC to include a measure of turning movements. Thus at a four-arm junction, surveyors would record both the flow and the direction of turn, see Figure 2.9. This sort of data would typically be used to analyse the traffic conflicts at a junction, so as to determine whether or not the junction needed to be modified.

Once again the count can be classified to identify the mix of traffic. This can be very useful as different types of vehicle have different acceleration, turning and braking characteristics, which will affect the amount of traffic that can pass through a junction.

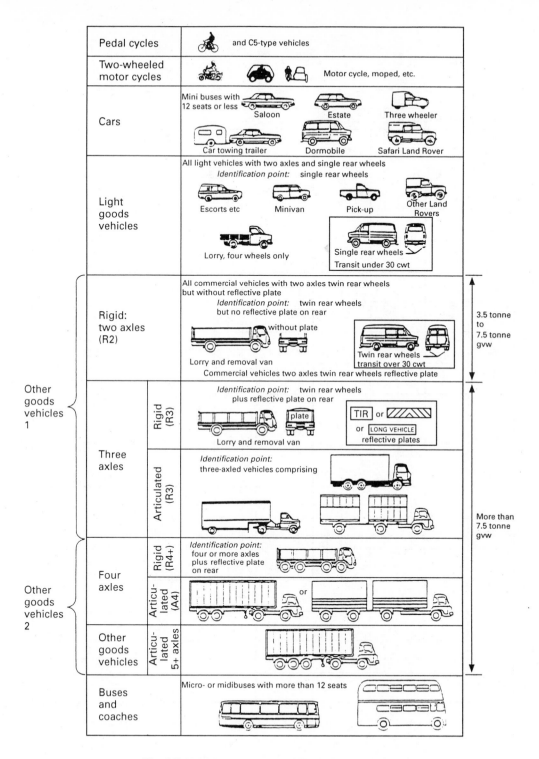

Fig. 2.7 Vehicle categories used for survey purposes.

Enumerator's name ....

Form No. `7` `7` COUNT POINT No. .... [| |] DATE .... DAY .... [ ] MINUTES COUNTED .... [|] (12 or 15)

Fields positions: 1 2 | 3 5 | 6 9 | 10 | 11 12

LOCATION ....

COUNT DIRECTION TOWARDS .... 13 [ ]

HOUR BEGINNING Delete those not applicable	QUARTER BEGINNING	CAR incl: estate car	TAXI	LIGHT GOODS 2 axles 4 tyres	HEAVY GOODS 2 axles, 6 tyres 3 axles or more	MOTOR CYCLES incl: scooter and moped	BUSES and COACHES	TOTAL
14 , 15	16 , 17	18 21	22 25	26 29	30 33	34 37	38 41	42 45
00	00							
04	15							
08								
12	30							
16								
20	45							
01	00							
05	15							
09								
13	30							
17								
21	45							
02	00							
06	15							
10								
14	30							
18								
22	45							
03	00							
07	15							
11								
15	30							
19								
23	45							
TOTAL								

Fig. 2.8 Manual classified count form.

Location diagram:			Enumerator				Job No:			Time start:	
			Date:				Location:			Time finish:	
			Weather:				Day:			Comments:	
	Turning movement / Vehicle type										
	Pedal cyclist										
	Motor cyclist										
	Cars/ LGVs										
	HGV 1 (2–3 Axles)										
	HGV 2 (4 Axles)										
	All bus types (inc mini- buses)										

Fig. 2.9 Turning count form.

2.4 Area-wide surveys

The surveys described above are adequate for measuring traffic flow and direction of movement at a single point, or at a single junction. However, if we wish to understand movement over a wider area, then other methods have to be used. Three techniques are described below, one for numberplate surveys and two covering origin and destination surveys.

2.4.1 NUMBERPLATE SURVEYS

We may wish to understand how traffic is circulating in a limited area. This could be, for example, a complex gyratory system, a residential area where we suspect that there may be rat-runs, or even a town centre ring road where we wish to understand if traffic uses the ring road or passes through the town centre.

The technique used is to record the registration mark of each vehicle as it enters and leaves the system being studied and then to match the registration marks, to establish how a vehicle travelled through the road system being studied. It is not normally necessary to record the full registration mark.

If the first four characters of a typical UK registration plate, which is usually of the form A 123 BCD are considered, then the probability of having the same four first characters on a different vehicle is the product of the probability of these same four characters appearing. There are 26 letters in the alphabet, although I, O and Z are not used, so there is a 1 in 23 chance of the first character being repeated. However, as not all the letters have not yet been used, the odds are slightly more, at present. Similarly there is a 1 in 10 chance of a number being repeated. In total, the chances of having two different vehicles appearing with the same registration number are

$$1 \text{ in } (23 \times 10 \times 10 \times 10) = 1 \text{ in } 23\,000$$

This risk is further reduced by the chance of the vehicles being the same type.

It can be seen from the above that, in practice, it is sufficient to record the first four characters of each registration mark as, in all but the largest systems, the chances of getting two or more vehicles with the same four first characters are insignificant.

In theory this is a very simple and robust survey technique. In reality, it suffers from a number of practical problems. The first of these is survey error. Even on the best run survey it is likely that 5–10% of registration numbers will be unmatched, due to errors in reading or recording number plates. Also, in larger traffic systems, vehicles can stop or start within the system and so remain unmatched. In difficult conditions, the level of mismatch can reach 30%.

For a complex situation, for example where there is more than one route between an entry and exit point, the survey could require data to be recorded at an intermediate point, as well as the entry and exit points. This adds to both the complexity of data collection and matching and analysis.

Obviously a vehicle cannot leave before it arrives, and to help prevent spurious matches the time that a vehicle is observed should also be recorded. This data also provides approximate journey-time information.

When the survey covers an extended area, for example a rat-running survey, then vehicles, such as those belonging to local residents may enter the survey area and stop, or start within the survey area. A vehicle may also enter the study area, stop for a while and then leave.

To ensure that one understands what the survey results represent, the survey has to be carefully specified to take account of these factors. Thus, for example, if a large number of vehicles are expected to leave and/or join the traffic flow within the survey cordon, high levels of mismatch can be expected. If however the system is closed, for example a gyratory system, then there should be a very high match.

2.4.2 ORIGIN AND DESTINATION SURVEYS

The alternative way to establish where drivers are travelling is to ask them, using an origin and destination (O&D) survey. Various types of O&D surveys are used as a part of the wider transport planning process. However, this is beyond the scope of this book and is not explored here. The standard techniques are roadside interview surveys and self-completion questionnaires.

In most cases it will be impossible to carry out a 100% survey of drivers and so we must rely on a response from a sample of drivers in the traffic flow. Clearly, if the survey results are to be relied on, the sample should be unbiased with all types of vehicles and movements represented.

2.4.3 ROADSIDE INTERVIEW SURVEYS

At a roadside interview survey, a sample of drivers are stopped at the side of the road and asked their origin and destination, plus any other data which could be of relevance, such as journey purpose. Figure 2.10 shows a typical survey form.

The size of sample will depend on flow and the level of reliability required. This is described in greater detail in Traffic Advisory Leaflet TA 11/81.[2] However, the theoretical advice offered in this guidance has to be balanced by what can be practically achieved. If an interview lasts for just a minute and, after allowing time for the driver to enter and leave the interview bay the time stopped is, say, 2 minutes, then each interviewer could handle 30 drivers an hour. Simple logic dictates that there has to be a limit on the length of an interview bay, for practical reasons, if not as a result of the road's geometry, and this will determine the absolute number of drivers that can be interviewed each hour.

Typical designs for interview stations are shown in Figure 2.11. Where surveyors are having to work close to moving traffic, the safety of all involved is paramount.

Traffic flow is directed past the interview point and a sample of vehicles is directed into the interview bays where the drivers can be asked about their journey. The power to direct traffic resides only with the police and so these types of surveys require the cooperation and continuous presence of a police officer.

As the direction of traffic at a survey station requires the presence of a police officer, it is important to involve the police in the design of the survey, to ensure that they are satisfied that the survey can be conducted safely and that the officer(s) involved are aware of the need to gain a representative sample from the traffic flow.

Once a driver has been selected for interview and is stopped in the interview bay, he should be asked to provide the necessary answers and then released as soon as possible. Although a driver must stop when instructed to do so by a police officer, there is no obligation on the driver to participate in the survey and a driver may refuse to answer any questions.

The data are used to reconstruct the pattern of vehicular movement by aggregating trip origins and destinations into a pattern of zones and then grouping together the trips to construct a matrix of movements called an origin/destination matrix. The exact grouping of information will depend on the road network and the distribution of developments served by the network. Obviously the boundaries of zones must be chosen so that, as far as possible, trip movements can be distributed correctly on the network.

As the interviews represent a sample of traffic, the survey responses have to be factored up to represent the total flow at the survey point. Normally this is done by undertaking a

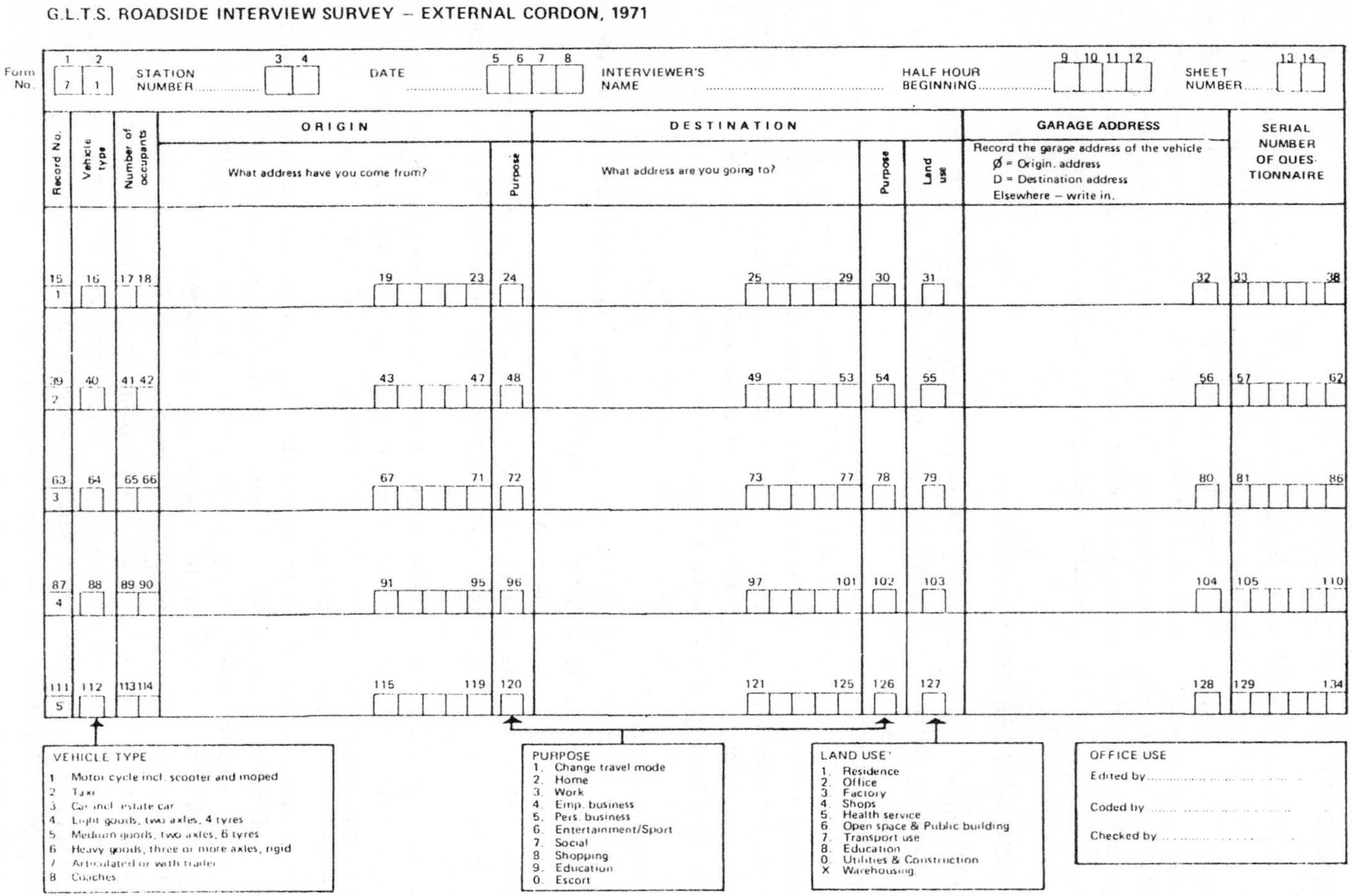
G.L.T.S. ROADSIDE INTERVIEW SURVEY – EXTERNAL CORDON, 1971

Form No. 7 1 | STATION NUMBER .... | DATE .... | INTERVIEWER'S NAME .... | HALF HOUR BEGINNING .... | SHEET NUMBER ....

Record No.	Vehicle type	Number of occupants	ORIGIN: What address have you come from?	Purpose	DESTINATION: What address are you going to?	Purpose	Land use	GARAGE ADDRESS: Record the garage address of the vehicle Ø = Origin. address D = Destination address Elsewhere – write in.	SERIAL NUMBER OF QUES-TIONNAIRE
15 / 1	16	17 18	19 – 23	24	25 – 29	30	31	32	33 – 38
39 / 2	40	41 42	43 – 47	48	49 – 53	54	55	56	57 – 62
63 / 3	64	65 66	67 – 71	72	73 – 77	78	79	80	81 – 86
87 / 4	88	89 90	91 – 95	96	97 – 101	102	103	104	105 – 110
111 / 5	112	113 114	115 – 119	120	121 – 125	126	127	128	129 – 134

VEHICLE TYPE
1 Motor cycle incl. scooter and moped
2 Taxi
3 Car incl. estate car
4 Light goods, two axles, 4 tyres
5 Medium goods, two axles, 6 tyres
6 Heavy goods, three or more axles, rigid
7 Articulated or with trailer
8 Coaches

PURPOSE
1. Change travel mode
2. Home
3. Work
4. Emp. business
5. Pers. business
6. Entertainment/Sport
7. Social
8. Shopping
9. Education
0. Escort

LAND USE
1. Residence
2. Office
3. Factory
4. Shops
5. Health service
6. Open space & Public building
7. Transport use
8. Education
0. Utilities & Construction
X Warehousing.

OFFICE USE
Edited by ....
Coded by ....
Checked by ....

Fig. 2.10 Roadside interview form.

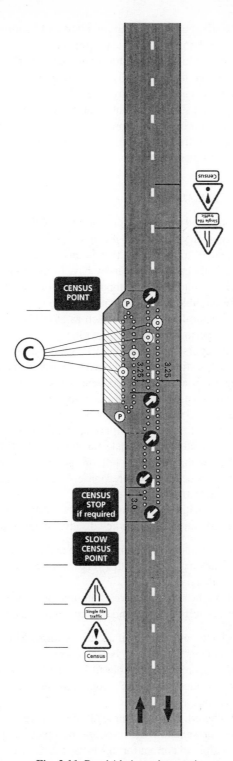

Fig. 2.11 Roadside interview station.

contemporaneous classified traffic count at the survey point and factoring up the sample results to the total flows. This process is known as sample expansion.

Sample expansion is normally achieved by factoring the sample in a given time period, say an hour or 30 minutes, to the observed flow, normally subdivided by vehicle type. Alternative time periods, such as morning and/or evening peak may be used.

It may be difficult to carry out a survey without so disrupting traffic that a contemporaneous count will be unrepresentative of normal traffic conditions. In these circumstances, the pragmatic approach would be to carry out a traffic count on the same day in the preceding or following week and to factor the survey results to these counts.

2.4.4   SELF-COMPLETION FORMS

In some locations, often in congested urban areas, it is not practical to set up an interview station. This could be because road geometry means that it is not possible to safely slow-down and stop traffic, or because of the volume of traffic means that an unacceptable level of traffic congestion would arise if road space were allocated for a survey bay. In these circumstances a reply-paid questionnaire may offer a suitable alternative methodology, to provide the information required.

In most places it should be possible to find a location where traffic flow can be stopped for a short period. This could be a natural interruption, such as at traffic signals, or as a result of a police officer stopping the flow periodically for a short time.

If traffic can be stopped, surveyors can enter the traffic stream and hand out a reply-paid card which asks the driver the same questions as would have been asked at a roadside interview. The card can be completed by the driver at their own convenience and posted back to the organisers.

Because the surveyors do not have to ask questions, many more cards can be distributed using the same resources as would be used for an equivalent interview survey.

Where there are multiple lanes of traffic it is important to ensure that the distribution of questionnaires between lanes is balanced. The rationale for this is self-evident. If there were three lanes of traffic at a survey point, with one turning left, one going ahead and one going right, then any bias in distribution could produce a bias in response, leading to a misrepresentation and a misunderstanding of the existing traffic flow.

The technique allows contact with a larger proportion of drivers in the traffic stream than would be possible with a roadside interview. The key disadvantages of the method are:

- The response rate cannot be judged in advance and can be highly variable.
- The sample size can vary greatly, with a very low response in some time periods and higher returns in others.
- The sample may not be representative of all time periods and vehicle types.
- The respondents are self-selecting and this may introduce a bias.

With the simple example of a three-lane road, if no survey forms were handed out in the right-turning lane then these movements would not be represented and, once the data had been processed and analysed it would prove impossible to use the survey results to reproduce the observed situation. If the issue of questionnaires is monitored and controlled, however, this situation can be monitored and the bias avoided.

If the questionnaires are numbered, then the sample response in each time period can be judged. If flow in one period is underrepresented in the response, or a particular category of flow is not fully represented, then it is possible to correct for this lack of data by a process of data patching. In simple terms this means that where there is an inadequate sample in one time period, the data from adjoining time periods are combined to allow representation of the traffic movement.

Data patching should only be attempted within carefully defined limits when traffic characteristics can be expected to be similar. For example, it would be acceptable to match successive peak periods when traffic flow is dominated by the journey to work. However, it would be wrong to merge peak and off-peak traffic.

The response rate to such surveys can vary tremendously and unpredictably. Responses can be as low as below 20% and above 50% for surveys which are similar in terms of the type of questions asked and the purpose of the survey.

It is now commonplace to offer an incentive to drivers to encourage a higher response rate. This would typically be a prize draw for a cash sum, or a free holiday or gift. There is no conclusive evidence that incentives regularly result in a higher response rate. However, there is no evidence that they deter responses and, on balance, they would seem to be more likely to help than hinder.

As with an interview survey, it is essential to carry out a traffic count so that the results can be factored up. Although the survey technique is designed to have the minimum impact on traffic flow, it is likely that there will be effects and so it is probably best to plan to record unobstructed traffic flows on another day.

The potential for this type of survey to disrupt traffic was graphically illustrated by a survey at traffic signals in West London on the A4, designed to capture traffic using the M4. The peak hour survey had surveyors handing out questionnaires to drivers at a signal stop line when the lights were red. The signal settings were unchanged and the extra delay to drivers was caused by a police officer who held the traffic on red/amber to ensure that surveyors were clear of the traffic. The survey was abandoned after about an hour, by which time there was a 20 km tailback on the motorway. In planning any survey, it is important to ensure that the planning takes account of the likely effects on traffic and seeks to minimise any adverse effects.

Occasionally it is not possible to survey traffic at the point where the information is required. With the benefit of hindsight, the A4 survey mentioned above was one such place. In these circumstances, a more time-consuming and expensive approach has to be adopted. For example, if an engineer wishes to understand the flow on a motorway link, it is not possible to set up a survey on the motorway. The technique adopted in these circumstances is to set up a series of interview stations on motorway accesses upstream of the part of the motorway which is of interest.

Self-completion forms can also be used to provide origin–destination information on bus passengers; the forms are distributed on the bus by surveyors. Once again, it is important to record both total passenger numbers and the time when each form was handed out, so that replies can be factored up to represent the full travelling population.

2.5 Speed surveys

There are three basic techniques for measuring the speed of traffic. The first method uses speed-measuring equipment, such as a radar gun, to record the speed of traffic, or a sample of traffic passing a particular point in space. The second relies on a vehicle travelling in the traffic flow, where the speed is calculated as the time taken to travel a certain distance. The third measures the speed of a journey, or the time taken to travel between two locations.

The first of these measurements is called the 'spot speed' for an individual vehicle. Spot speed measurements can be used in combination to show the variation of vehicular speeds, as a simple frequency graph, see Figure 2.12. Alternatively spot speed measurement can be used to be use to calculate the 'time mean speed' of traffic passing the measuring point.

Time mean speed is the average speed of vehicles passing a point over a specified time period and is defined as

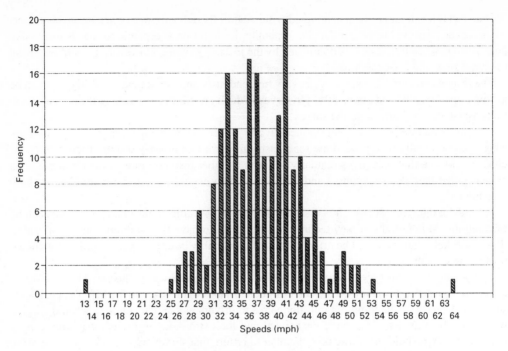

Fig 2.12 Spot mean speed.

$$V = \sum_t V_t / n$$

where

> V is the time mean speed;
> V_t is the speed of an individual vehicle;
> n is the number of vehicles observed.

The data required can be collected using a speed measurement device, such as a radar gun, suitably positioned to take readings of the traffic stream it is desired to study. This is illustrated in Figure 2.13.

The second measure of speed is 'space mean speed'. This is a measure of the speed of travel over a measured distance, rather than at a single location. Thus if an observer wishes to know the speed of vehicles travelling along a length of road length l, then if each vehicle i takes time t_i to travel the link then space mean speed is defined as

$$V = l / (\sum_i t_i / n)$$

where

> V is the space mean speed;
> t_i is the travel time of the ith vehicle;
> n is the number of vehicles observed.

The final measure of speed is 'journey speed' or, more correctly, journey time. Journey speed within a network is measured using what is known as the floating car method. The technique measures average journey time between two locations in a network, along a predetermined route.

Fig. 2.13 Hand-held digital radar gun.

With the floating car method, a car is driven at the average speed of the traffic. This is achieved by driving the car so that it overtakes as many cars as overtake it. Hence the name floating car, because the vehicle floats in the traffic stream, moving at the average speed of the traffic on the network.

Clearly, over an extended journey this may not be too difficult to achieve; however on a short link, or in a busy street, overtaking opportunities may be limited and so the technique may be impractical.

The technique allows an extended survey of speeds throughout a network, with limited survey resources and can be very useful in gaining a broad understanding of traffic speeds in an area. It should be noted that, where average traffic speeds exceed the speed limit this technique is, strictly speaking, inappropriate as the observers would have to speed to collect the data.

Journey speed surveys are often used to collect time series data on highway networks to show how traffic speeds are changing with time. Data is usually collected by recording the journey time in stages, between successive major junctions say, with predetermined timing points on the vehicle's journey. Obviously, time on a single link can vary for a number of reasons. These could include:

- random variations in traffic flow;
- time of day;
- the turning movement the vehicle makes at a junction;
- incidents, such as road works or an accident;
- for traffic signals, where the vehicle arrives in the signal's cycle.

To obtain an average value, it is essential to repeat the journeys a number of times to give data from a range of traffic conditions. Obviously any survey data collected where it can be established that the traffic flow was affected by an accident or road works should be discarded, unless of course the purpose of the survey was to measure the disruption effect of the incident. Pragmatically, the number of runs is likely to be influenced by the budget available for the survey, but a minimum of five good runs should be undertaken, although as few as three have been used in large scale surveys such as the Greater London Speed Surveys.[3]

The first two methodologies are appropriate for precise measurements at a single point, or in a single street. The third technique allows traffic speed to be measured across a larger area and would be used to collect time series data.

2.6   Queue length/junction delay surveys

Queue length surveys involve recording the length of the queue on an approach to a junction. Typically, the position of the back of the queue is recorded for each lane every 5 minutes at a roundabout and at signalised junctions approximately every 5 minutes as the maximum queue during the current cycle. A typical form of presentation of the resulting information is shown in Figure 2.14.

Junction delay surveys involve one surveyor recording the times and registration numbers of vehicles joining the end of the queue and a second surveyor recording the times and registration numbers of vehicles passing through the junction. By matching registration numbers, very accurate measures of average delay and the standard deviation of delay can be obtained.

2.7   Video surveys

The use of video as a data collection tool in traffic engineering is a relatively new but potentially very powerful concept. A strategically placed camera can be used to observe traffic and parking activity in a street and, depending on location and equipment, it is possible to survey up to 400 m of road from a single vantage point.

Cameras are mounted high to minimise the obstruction of the longer view from vehicles near to the camera. One of the key advantages of a camera is that it records everything that happens. Other survey techniques inevitably only record partial data, collecting just those aspects of traffic behaviour which the survey is designed to record. With a video survey, it is possible to review the video and observe other activities which were thought to be unimportant when the survey was planned.

The video has a particular advantage when flows are very high and it is difficult to count manually, or when we wish to study a particular location where we are not absolutely sure what the key issue is. A video can simultaneously record:

- traffic flow
- turning movements
- speeds
- congestion and delays
- parking and loading
- pedestrian movements.

Most importantly, a video shows the interaction of all these factors.

The video also offers the unique advantage of allowing us the opportunity to view the situation repeatedly, until we are satisfied we understand what is happening.

Video surveys are not cheap. Although the data collection may only require the presence of a

QUEUE LENGTH SURVEY

SITE:- **Stratford Road / Highgate Road**   DAY:- **Wednesday**

DIRECTION:- **Southbound**   DATE:- **22nd May 1996**

Key   Lane1   Inside/Nubside Lane
      Lane2   Offside Lane   - - - -

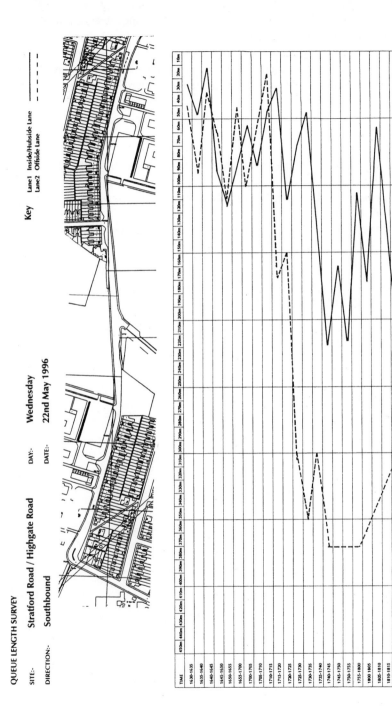

Fig. 2.14 Queue length/junction delay survey.

single technician, to monitor the equipment, the subsequent recording and analysis of data from the video can take up to six times as long as the real time recording, depending on what data is to be collated from the video and whether or not computer-assisted techniques are available. The authors use a system where the video is displayed under computer control so that an operator can respond to data-entry requirements and readily key information directly into a database. Video survey techniques employing numberplate matching with split-screen presentation of the data for the analyst can also be used as a substitute for manual methods of undertaking journey time and area-wide surveys.

References

1. Institution of Highways and Transportation & Department of Transport (1987) *Roads and Traffic in Urban Areas*, HMSO, London.
2. Department of Transport (1981) *Traffic Surveys by Roadside Interview*, Traffic Advisory Leaflet 11/81, HMSO, London.
3. Greater London Council (1967–1986) *Greater London Speed Surveys*, Greater London Council.

3
Parking Surveys

3.1 Introduction

Every trip by a vehicle results in a parking act at the end of the trip. The vehicle may be parked on the street or off-street in a car/lorry/cycle park. To help understand parking behaviour, there are a multitude of parking survey techniques that have been developed, each aimed at measuring something slightly different. Each technique is discussed below, in terms of the order of complexity.

3.2 Supply surveys

In order to understand existing parking behaviour, and the potential for accommodating additional parking, it is essential to have a sensible estimate of the amount of parking. This is not always as simple as it may sound, because cars and other vehicles can be parked in many different places.

On the street, in controlled parking areas, street parking is explicitly marked either as individual parking bays, or as lengths of street where parking is allowed. The bays can be counted explicitly: for lengths of road where parking is allowed, it is reasonable to use an estimate of 5 m of kerb space for each car parking space. When undertaking a survey of the spaces available on-street, it is important to remember that restrictions may only apply part of the time. This means that the supply of available parking space could vary, according to the time of day.

Off the street, land and structures that are designed to be used as parking are usually marked out with parking spaces which can be counted. However, it is commonplace to see yards, service roads and other areas used for parking on a regular basis, and these can make a significant contribution to total parking supply. In Central London, a 1977 Census survey, where all the places regularly used for car parking were recorded, identified some 57 000 car parking spaces. A later review of planning permissions showed that there were about 34 000 spaces with planning permission, a 40% difference.

3.3 Occupancy surveys

The simplest parking survey is an occupancy survey, where the number of vehicles parked on a street, in a car park or parking area are periodically counted. A surveyor passes round the parking spaces at predetermined intervals and simply counts the number of vehicles in the parking place. The surveyor may record:

- the total number of vehicles;
- the number of vehicles in each street or length of street;
- the numbers of each type of vehicle, by street or street length.

The survey technique tells us little about the vehicles, in terms of their arrival, departure and duration of stay; however, it does allow us to gauge the adequacy of the car parking available, when compared with parking demand. The surveys also tell us how busy the parking is at different times of the day or week.

This survey technique is appropriate where the data is being collected to give either a broad understanding of the adequacy of the parking supply or an understanding of changes in demand over time. For example, if a car park operator wished to ensure that he always had enough parking available to be sure that a driver could always expect to find a place to park, he might set a threshold of 85% occupancy and, when demand reached this level, he would either increase supply or make the parking less attractive, for example by raising charges, so that demand was kept below the threshold.

The technique allows a large amount of parking to be surveyed quickly and so requires fewer survey resources than the other, more precise, methods described below.

3.4   Beat surveys

To obtain more detail about the behaviour of individual vehicles, and hence an aggregate picture of parking behaviour, a beat survey can be used. In a beat survey the surveyor visits, in turn, a predetermined number of parking spaces and records details of the vehicles that are observed parking in each space. Typically the surveyor would record:

- time
- parking space location
- vehicle type
- partial vehicle registration number (discussed in Chapter 2).

Normally a beat survey is undertaken at regular intervals and so the time is recorded to an appropriate time block. Thus if the survey were hourly, the time would be recorded as the hour in which the survey round took place, and so on. Figure 3.1 shows an example of a survey form for a street survey of a typical area.

The beat frequency will be determined by the purpose of the survey. If an area were used mostly by residents, who tend to park all day, or by workers who arrive in he morning and leave at the end of the working day, then a survey may only be required every 2 hours. However, if the survey were in a high street, where vehicles are coming and going every few minutes then a 15 minute beat would be more appropriate.

Such a survey could also be used to identify vehicles that regularly park in the same place, for example commuters who park in the same street every day. In this case the survey could take place just once a day, over say a week.

More commonly, this survey technique is used to understand patterns of arrival and departure and duration of stay. It distinguishes between all-day and short-stay parking activity. Three or four visits a day would allow an unambiguous quantification of long-stay parking; however, it could considerably under-count short-stay numbers (discussed in greater detail later).

Where it is important to identify short-stay parking, then the beat frequency needs to be much higher, possibly as often as four times an hour. Even this can under-count very short duration acts. The methods that can be adopted to address this problem are discussed later.

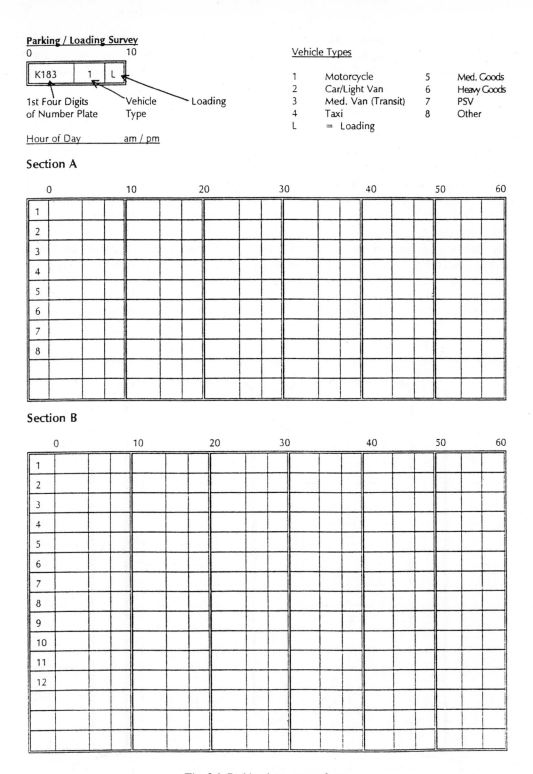

Fig. 3.1 Parking beat survey form.

If data are to be compared between succeeding beats it is very important that the parking act being observed can be located exactly to a particular parking space. If this is not done, we could have the situation where the act observed in space (a) on one pass of the surveyor is compared with an act observed in space (b) in the next round.

It is also very important to ensure that the surveyor passes round the beat in the same order at each visit, to avoid distortion of the results. The reason for this can be explained with a simple illustration. Figure 3.2 shows the path a surveyor is expected to follow round a car park for a 30 minute beat survey, starting at **S** and ending at **F**. If the survey takes 25 minutes to complete, with a 5 minute break at the end of each pass, then successive observations at each bay will be about 30 minutes apart. If, however, after completing a beat in the correct order the surveyor were to retrace his steps, visiting the parking bays in reverse order, then the two observations of the last bay would be only 5 minutes apart, whereas the two observations of the first bay would be separated by an hour.

Beat surveys can be used for surveys on the street or in a parking area. In either case careful planning is needed to ensure that each parking place can recognised, to allow correct comparison of data between successive beats.

In most car parks the bays are marked and so location can be explicitly identified, provided the same route round the car park is followed. In a street, where there may be no markings, location

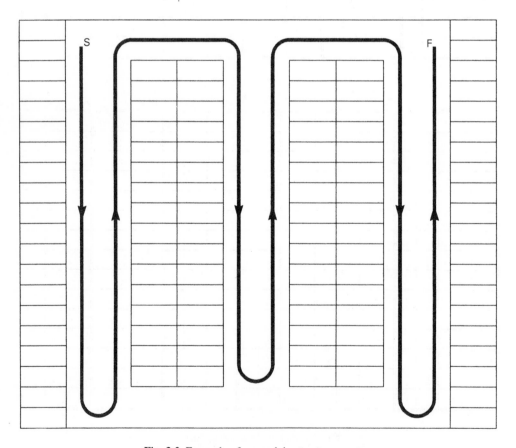

Fig. 3.2 Example of car park beat survey route.

is less clear and can be confusing. To avoid this, the street should be divided into nominal parking bays at the planning stage of the survey (Figure 3.3). On the day of the survey the bay limits should be marked on the kerb with chalk, or some other nonpermanent marker, to aid the surveyors.

Any parked vehicle should be allocated to the bay it most nearly sits in. This should not cause confusion as, if a vehicle straddles two bays, it should be allocated to the bay it occupies most of. Clearly until the first vehicle leaves, a second vehicle cannot occupy the same space so no confusion should occur.

If the purpose of the survey is to understand both the type and number of vehicles parked, then the type of vehicle seen should be recorded. Even if this information is not required, it can still be a useful way of checking data quality and helping to clarify uncertainties about the survey results.

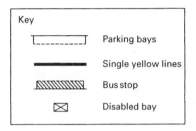

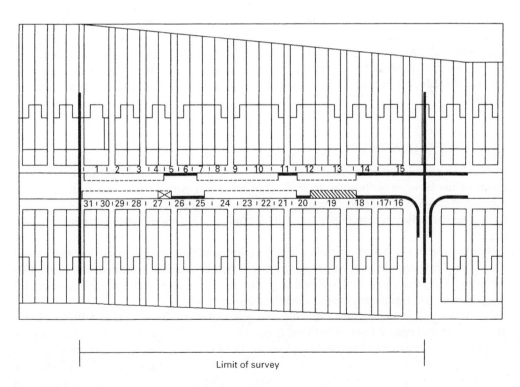

Fig. 3.3 Survey design for street parking.

The complexity of vehicle types recorded will depend on the use to be made of the survey results. A minimum might be cars and other vehicles; however, if one wishes to gain a more detailed understanding of activity, cars could be subdivided into car, taxi and orange badge holder, and other vehicles might be split into motorcycle, buses and various subcategories of commercial vehicle.

Whatever the categorisation adopted, it is clearly important to ensure that surveyors are fully and unambiguously briefed on how to identify the types of vehicle to be used for the particular survey.

Of course, if the purpose of the survey is to record the number of times a particular vehicle visits, then it would be necessary to record the full registration plate.

Beat surveys compare a series of snapshots of activity, and can explain the parking behaviour of the vehicles in the area being studied by identifying the changes between the snapshots. However, because the survey only observes the vehicles intermittently, they do not give a completely accurate picture of the behaviour being observed. The survey technique suffers from two main types of inaccuracy, as described below.

3.4.1   TIMING ACCURACY

When a vehicle is first seen on a beat the observer does not know exactly when it arrived, only that it arrived after his last visit and before the present one. Similarly, when the vehicle leaves the observer does not know the exact time of departure. It can be seen that if the beat frequency is once every t minutes then, at one extreme, the vehicle could have been parked for up to $2t$ minutes longer than the observer has logged. At the other extreme the vehicle could have arrived just as the observer reached the parking place on one pass and left just as he/she left the parking place on another pass, in which case there would have been a zero error on the time recorded. It follows that the average error of observation is t, the beat frequency, and so when calculating length of stay all observations should have this amount added to give an unbiased estimate of duration of stay, i.e. if a vehicle is recorded n times, then the best estimate of its duration of stay is $(n + 1)t$ minutes.

3.4.2   UNDERCOUNTING

With a beat survey a certain number of vehicles will arrive and depart between successive passes of the surveyor without being seen. Thus short-stay parking acts, that is those with a duration of stay less that the beat frequency, will always be undercounted on a beat survey. The scale of undercounting and its importance will depend upon the specification and purpose of the survey.

For example, in a long-stay car park, where the purpose of the survey is to measure the amount of long-stay parking, the fact that short-stay acts have been underrepresented will be of little importance. However, in a survey aimed at recording short-stay illegal parking in a restricted street, a beat could significantly undercount the number of acts.

This factor was first recognised explicitly as a potential major error in the mid 1970s when it became clear that central London suffered from extensive illegal parking and yet beat surveys were failing to satisfactorily explain this behaviour. In 1977 the GLC undertook a comprehensive survey of parking in central London.[1] As part of this study the vehicles observed in a 30 minute beat survey were compared with the actual number of vehicles parking. The results are shown in Table 3.1.

Table 3.1   Level of undercounting with a 30 minute beat survey

Parking type	Acts observed (%)
Two hour meters	89
Four hour meters	93
Residents' bays	90
Single yellow line	41

(Source: Carr *et al. The Central London Parking and Car Use Survey*[1])

Table 3.1 clearly shows that, although the undercounting on the longer stay (compared with the beat frequency) parking is around 10%, for the short-stay (illegal) parking the survey only recorded about two in every five acts. Clearly, if the beat data had been used as a basis for judging the level of illegal parking, the scale of the problem would have been massively understated and any conclusions based on the results would have been invalid.

Recognition of this deficiency led to the development of the continuous observation street survey techniques described below.

3.5   Continuous observation surveys

If we do not need to know where within a larger car park a particular car has parked, it may be more efficient to simply record the vehicles as they enter and leave the car park. In this case, the time of arrival, vehicle type and registration are recorded on entry and again on exit from the car park. The two data sets are then compared to establish how long each vehicle was parked.

This technique is more accurate than a beat survey as the duration of stay is known exactly, within the accuracy with which the data is recorded. With this type of survey, a large car park can be observed with just two or three people.

The data can be recorded on a form of the type shown in Figure 3.4. Alternatively at a very busy car park, the surveyor could record the data using a tape recorder, for later transcription and analysis. However, tape recorders should be used with care, as they introduce a whole different set of risks, such as flat batteries, reaching the end of the tape at an inopportune moment, or being unable to hear the surveyor over background noise.

The accuracy of an entry/exit survey in a car park can be transferred to observing parking in the street with a continuous observation survey. With this type of survey, each surveyor is limited to watching just a few parking places, as many as can be seen from one location. Typically this would be 30–40 spaces.

The surveyor is able to exactly record the arrival and departure time of each vehicle and, if appropriate, can also record other relevant information, such as whether or not the vehicle was loading, if the parking was paid for and, if so, when the payment expired. The surveyor can also record details such as enforcement activity, if this is required.

In 1982 the Transport Research Laboratory (TRL) developed a computer package specifically designed to undertake this type of survey, called PARC.[2,3] PARC is designed for use in a HUSKY HUNTER portable computer. The program allows the surveyor to record each detail of the parking activity being observed in real time. Subsequently, the data is transferred to a PC and the analysis package allows the data to be analysed and tabulated, to give standard summary results of the activity observed.

Car Park Name.......................... Site Number....... BUTP II PARK3

Surveyor's Name................... Date...........

Weather : dry/rain

Vehicle Registration	Type	Time In	Time Out

Vehicle Registration	Type	Time in	Time Out

Notes:

Vehicle Type: 1 car
2 taxi
3 two wheeler
4 other

Fig. 3.4 Example of continuous survey form.

3.6   Summary

The engineer wishing to survey parking supply or activity has a wide range of techniques to call on. The method chosen will depend upon the rationale for the survey.

References

1.   Carr, R, Baker, LLH and Potter, HS (1979) *The Central London Parking and Car Usage Survey*, Greater London Council.
2.   JMP Consultants Limited (1989) *PARC Suite Users Manual*, JMP, London.
3.   Transport and Road Research Laboratory (1991) *Guide to PARC.2P*, HMSO, London.

4

Estimating Travel Demand

4.1 Introduction

The estimation of travel demand is a fundamental part of traffic engineering design work. The key questions are how much effort needs to be expended in estimating demand and what method should be adopted. The answers depend on the nature of the design issues. For example, a minor traffic management design to improve road safety over a length of road in inner London where traffic flows have been stable for many years will require little more than a survey of existing traffic. The reverse is true of a proposal for a new roadway to assist regeneration in an old urban area, where design will depend on estimating the new traffic likely to be attracted to use the new road.

Estimation techniques fall into three main categories:

- growth factor
- low-cost manual estimation
- computer-based traffic models.

All of these techniques include assumptions about the four basic elements of estimation, which are:

- trip generation
- trip distribution
- modal split
- assignment.

A trip is defined as a one-way travel journey between the origin (start) and destination (end) of the journey. Trip generation is the number of trips starting or ending at an area (or zone) in a given time period, such as per day or hour.

Trip distribution describes the number or proportion of trips from an origin zone spread amongst all destination zones.

Modal split is the split (or share) of these trips among different modes of travel (e.g. car, public transport, walk, cycle).

Assignment is the process whereby trips are routed from their origins to their destinations through a travel network (e.g. road, bus and train routes, cycle ways/routes).

These four basic elements, shown in Figure 4.1, define the numbers of trips made from an area, the destination of these trips, the modes of travel used and the routes taken.

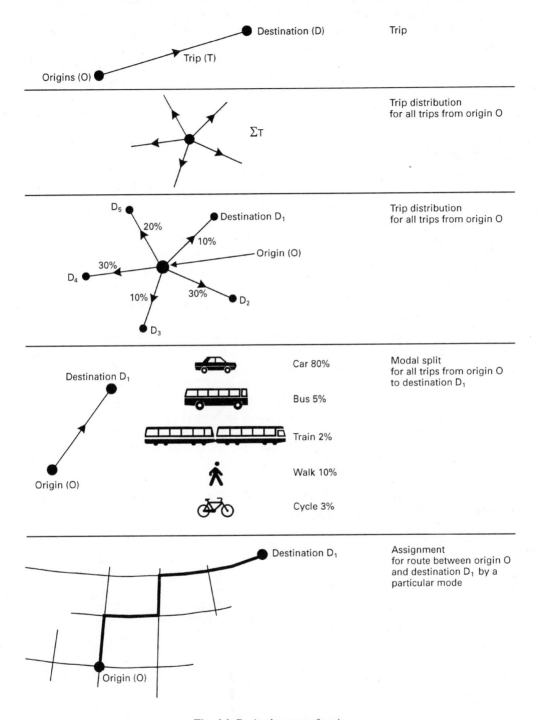

Fig. 4.1 Basic elements of a trip.

The choice of estimation technique will depend on the complexity of the demands to be estimated and the resources available for the estimation. For example, if an estimate of the future number of vehicles on a rural road is required, then a growth factor method may be appropriate. If the impact of a new housing development on the local road network is to be assessed then low-cost manual estimation is likely to be adequate. Estimation of travel demands by different modes of transport in an urban area will probably require the use of computer-based traffic models.

4.2   Growth factor

The Department of Transport has prepared National Road Traffic Forecasts[1] (NRTF) which are based on national models with current policies, the best available evidence of behaviour and the capacity of the current road network. The forecasts are of increased vehicle kilometrage, which translates directly into an increase in traffic on a section of road.

The forecasts have been prepared to year 2031 as shown in Table 4.1. As an example, the forecasts predict a growth in all traffic between 1996 and 2011 of 28%. The growth over these years varies by road type, for example growth on all urban roads is estimated at 25%, all rural roads at 30% and rural motorways at 52%.

Use of these forecasts is appropriate only when counts of traffic are available for the particular section of road being studied. When the origins and destinations of trips on the road have been surveyed (or are available from a traffic model) then local growth factors rather than the national forecasts can be used. These local growth factors are also based on the national models and are available at local authority district level in these models. At this disaggregated level, they reflect local zonal projections of planning data. The local factors are available as trip-end estimates for each district from the Department of Transport's TEMPRO program. A national forecast adjustment factor (NFAF) then has to be applied to make these local forecasts of travel compatible with the NRTF, which records the forecast growth in vehicle kilometres. In its simplest form, this approach can give an estimate of future growth as the average of the local factors weighted by the trips from each origin and then adjusted by NFAF.

This growth factor approach then estimates the future trip generation and distribution, while assuming that modal split and assignment remain constant.

Table 4.1 Traffic growth

Year	Central estimate for traffic growth for all vehicles
1996	100
2001	109
2006	119
2011	128
2016	138
2021	146
2026	153
2031	160

4.3 Low-cost manual estimation

The starting point for low-cost manual estimation will normally be an origin–destination (O–D) matrix of traffic. This matrix could be obtained from a survey of traffic using a section of road, normally by roadside interview, and this would be appropriate if a local improvement to part of the road network is proposed.

Alternatively, if expansion of existing developments or new developments is planned, then a survey of existing residents, employees and visitors or of residents, employees and visitors at nearby similar developments can provide the origin–destination information required.

The matrix will be growthed to the forecast future year by using trip-end growth factors, that is factors applied to the origins and destinations of the matrix. These factors can be the local growth factors by district as previously discussed, locally calculated factors at a more disaggregate level, or values forecast for new developments as described in Chapter 13. There are several simple matrix-manipulation techniques such as the Furness technique for applying these growth factors to the O–D matrix.

The Furness technique involves an iterative balancing of the rows and columns of the matrix so that the growth in origin and destination trip ends, as well as the new total flow, represented by the total of the matrix elements, are correctly represented as follows:

For

T_{ij} trips between origin zone i and destination zone j surveyed
O_i trip end for origin i surveyed
D_j trip end for destination j surveyed

and the NFAF-adjusted growth factors

g_i for origins
g_j for destinations

then

new origin trip end for zone i	$=$	$O_i g_i$
new destination trip end for zone j	$=$	$D_j g_j$
new total flow	$=$	$\sum O_i g_i$

The next step is to assign the matrix to the road network. In its simplest form of representation, the network will consist of road links with measured or estimated journey times and lengths. A simple network representation is shown in Figure 4.2. The assignment process typically involves identifying the quickest route from each origin to each destination, based on journey time, or the shortest route, based on distance.

Traffic between origins and destinations is then loaded onto the network. For a small network (e.g. up to 50 links) and a small matrix (e.g. up to some 50 elements) then manual loading may be practical. For more links and elements it will be necessary to consider the use of automatic network assignment techniques described in the next section.

A more sophisticated approach than manual assignment is to use a diversion curve.[2] This may be appropriate when two competing routes are available, for example when a local bypass is being considered. The approach allows for multiple routeing to represent the different perceptions of best route that drivers have, and provides a more accurate estimate of the proportion of existing traffic that will switch to using a bypass.

When considering new or expanded developments, forecasting is undertaken in two parts: traffic from the new developments is estimated based on O–D data and assigned to the network;

existing traffic on each link is then growthed to represent future years using the factors described and then added to the traffic from the new developments. This is described in more detail in Chapter 13.

The low-cost manual estimation then adds a manual assignment to trip generation and distribution but assumes that the modal split remains constant.

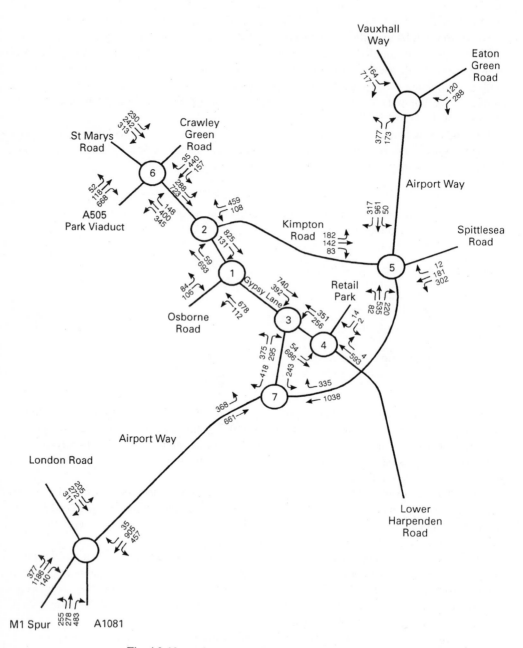

Fig. 4.2 Network representation showing traffic flows.

4.4   Computer-based traffic models

A traffic model is a set of mathematical equations which, when taken together, provide an estimation of traffic flows. Some models are designed to investigate strategic planning issues at a broad level, to assess specific policy options, such as congestion charging, or to identify the interaction between land use changes and transport. These models are unlikely to be used directly by traffic engineers and so this section deals only with serial models, such as SATURN[3] and TRIPS,[4] which provide the four basic elements of estimation:

- trip generation
- trip distribution
- modal split
- assignment.

Typically the model will be disaggregated by time of day (morning peak, inter peak and evening peak). The trip generation or trip-end model may also be disaggregated by purpose and vehicle type:

- person home-based work trips;
- person home-based education trips;
- person home-based shopping trips;
- person home-based other (e.g. leisure and social) trips;
- person non-home-based (e.g. employer's business) trips;
- commercial vehicle trips;
- other vehicle trips (e.g. bus, coach and taxi).

The trip-end model will be based on relationships involving population, employment, car ownership and land-usage characteristics, and is often determined using linear regression.

Trip distribution is often based on a gravity model employing the concept of generalised cost of travel. Generalised cost C puts a monetary value on the distance travelled and the time spent travelling and includes other travel costs, so that:

$$C = a \times \text{travel time} + b \ \times \text{travel distance} + d$$

where

a	is the value of time as assessed by the Department of Transport;[5]
b	is the driver's perceived vehicle operating cost;[5]
d	is other travel costs, e.g. parking and toll charges.

The common form of the gravity model is:

$$T_{ij} = O_i D_j f(C_{ij})$$

where T_{ij}, O_i and D_j are as described on page 39 and

$$f(C_{ij}) = \alpha \exp(-\beta C_{ij})$$

with α and β being estimated parameters.

Typically α enables the resultant matrix to balance to correctly represent total traffic flow and β is calibrated against surveyed flows. Where calibration is not possible, then the deterrence function is often simplified by traffic engineers to

$$f(C_{ij}) = \frac{\alpha}{C^2_{ij}}$$

which is comparable with Newton's gravity model.

In urban areas, it is now common to asses the modal share of travel, i.e. the proportion of travel by private transport, public transport, walk and cycle. The logit model is often used and is of the general form:

$$T_{ijk} = \frac{T_{ij}\exp(-\lambda_k C_{ijk} + \delta_k)}{\Sigma_n \exp(-\lambda_n C_{ijn} + \delta_n)}$$

where

T_{ijk}	is the number of trips between zones i and j by mode k, for example public transport;
T_{ij}	is the number of trips between zones i and j by all modes;
n	is the number of modes being considered;
λ_k, λ_n	are cost parameters for modes k or n;
C_{ijk}, C_{ijn}	are the generalised costs of travel by modes k or n;
δ_k, δ_n	is the modal handicap, for example the monetary value travellers place on being able to use private transport rather than public transport.

Trip assignment models in basic form identify the least-cost route through a transport network between origin and destination and then load the estimated O–D matrix for a travel mode onto that network. This basic form is known as all or nothing, because all traffic is assumed to select only one route for a journey between an origin and a destination.

Two refinements are commonly adopted; these are stochastic assignment and capacity restraint assignment. Stochastic assignment takes account of the different perceptions that drivers have of the best route and so spreads traffic over more than one route. Capacity restraint assignment allows for the increase in travel time caused by congestion and the resultant decision by drivers to use a different route to their preferred route. This is often achieved using an equilibrium approach which has, as its basis, that traffic on a congested network arranges itself such that all routes used between an origin and a destination have equal minimum costs.

A standard procedure in developing a traffic model is to first calibrate and validate it before it can be used for forecasting travel demand. Calibration and validation requires the development of observed trip O–D matrices.

Ideally, these matrices are based on surveyed travel data. However, it is not always possible to obtain a full coverage of an urban area and so two techniques are commonly used, partial matrix and matrix estimation. The partial matrix technique involves filling-in missing parts of the matrix using the travel data that is known. The matrix estimation technique uses the method of maximum likelihood to estimate the matrix using road link counts; this technique can also be used to update a matrix obtained in the past.

Once the model parameters have been established by calibration against the observed O–D matrices, then an independent set of traffic data, typically counts on screenlines, is used to validate the accuracy of the model. The model is then ready for the traffic engineer to use when forecasting future travel demand for a set of planning assumptions on future land use location and transport infrastructure.

4.5   Accurate and appropriate data

However good the technique chosen for estimating travel demand, its value depends on the accuracy and appropriateness of the traffic data used. The main issues in determining the amount and type of data to be collected are the time period for appraisal and the purpose of the demand estimation.

The Department of Transport (DoT)[6] requires that daily traffic levels are established for trunk road appraisal. This means that any roadside interview surveys should cover at least 12 hours of the day to provide a reasonably accurate daily O–D matrix. Automatic traffic counts are typically used to establish annual average daily traffic (AADT) levels and to provide the factor to convert 12 hour survey information to represent 24 hours.

The DoT[7] also requires that junction merge and diverge and weaving section calculations be based on a percentile highest hour flow. Roads are classified into three types, based on a seasonal coefficient with the percentile highest hour flow to be estimated for demand and appraisal purposes shown in Table 4.2. Automatic traffic counts can provide the percentile highest hour flow.

Table 4.2 Trunk road types

Road type	Seasonal coefficient	Percentile highest hour flow
Main urban	1.05 <	30th
Interurban	1.05–1.25	50th
Recreational/interurban	> 1.25	200th

The seasonal coefficient is given by

$$\text{seasonal coefficient} = \frac{\text{flow in August}}{\text{flow in a neutral month}}$$

Neutral months are taken as April, May, June, September and October, and it is these months that are normally selected for surveys on trunk roads and also nontrunk (county) rural roads.

Peak period analyses are normally required for nontrunk urban roads. Surveys must then cover the peak periods and identify the profile of flow within these periods. Typically, a highest flow over one hour, known as the peak hour which can vary from site to site, is identified and used for analysis. Taking a turning movement count at an urban junction over a single day will give a measure of the highest flow. Its accuracy as representative of average peak hour conditions over the year will be relatively low because of variations of flow from day to day. This accuracy can be enhanced by adjusting the single-day count to represent average flows obtained from automatic traffic counts over a longer period. Any attempt to assess average peak hour conditions must allow for this variation in flow that occurs naturally.

Individual counts are themselves subject to errors that are not related to this variation. These errors can occur from three sources:

- data sampling errors
- measurement errors
- human errors.

Sampling errors typically occur in surveys. For example, a roadside interview survey will often only achieve an interview sample of some 25% of drivers passing the interview point; these 25% may be unrepresentative of the total. The value of the sampling error can be estimated if reasonable assumptions are made about the way in which the sample is drawn.

Measurement errors depend on the method of measurement. Studies have shown that an automatic traffic count of one day will be accurate to within ±5% of the actual count to a 95% confidence interval. Manual classified counts give accuracies of within ±10% for all vehicles and ±18% for all goods vehicles, again within a 95% confidence interval.

Human errors occur when data is processed and analysed. Quality checking procedures are needed to minimise them.

The forecasting process itself introduces further errors as any predictive model relies on assumptions that cannot be proven in advance. Estimates of travel demand then must always be considered as only that, i.e. estimates, and any analyses undertaken on the basis of estimated travel demand should be subject to sensitivity tests. These tests should cover a range around the estimated demand to ensure that any traffic engineering decisions on implementing physical changes or introducing new traffic regulations are robust against inaccuracies in the estimated demand.

References

1.   Department of Transport (1997) *National Road Traffic Forecasts (Great Britain) 1997*, HMSO, London.
2.   Department of Transport (1996) *Design Manual for Roads and Bridges*, Volume 12, Traffic Appraisal of Roads Schemes, HMSO, London.
3.   Leeds, ITS and Atkins, WS (1995) *SATURN 9.2 User Manual*, Leeds.
4.   MVA (1994) *Introduction to TRIPS Version 7*, MVA, Woking.
5.   Department of Transport (1996) *Design Manual for Roads and Bridges*, Volume 13 Section 2, Highway Economics Note 2, HMSO, London.
6.   Department of Transport (1996) *Design Manual for Roads and Bridges*, Volume 13 Section 1, Economic Assessment of Roads Schemes (COBA), HMSO, London.
7.   Department of Transport (1992) *Layout of Grade Separated Junctions* HA 22/92, HMSO, London.

5

Capacity Analyses

5.1 Capacity definition

The term capacity when referring to a highway link or junction is its ability to carry, accommo-date or handle traffic flow. Traditionally, capacity has been expressed in numbers of vehicles or passenger car units (PCU). (Vehicles vary in their performance and the amount of road space they occupy. The basic unit is the passenger car and other vehicles are counted as their PCU equivalent, so that a bus might be 3 PCU and a pedal cycle 0.1 PCU.) In recent years public transport operators have applied pressure to consider highways in terms of their passenger handling capacity and thus give a greater emphasis to the benefits of using high-occupancy vehicles such as buses or trams.

There is no absolute capacity value that can be applied to a given highway link, traffic lane or junction. The maximum traffic handling capacity of a highway depends upon many factors including:

- The highway layout including its width, vertical and horizontal alignment the frontage land uses, frequency of junctions and accesses and pedestrian crossings.
- Quality of the road surface, clarity of road marking; signing and maintenance.
- Proportions of each vehicle type in the traffic flow and their general levels of design, performance and maintenance.
- The numbers and speed of vehicles and the numbers of other road users such as cyclists and pedestrians.
- Ambient conditions including time of day, weather and visibility.
- Road user levels of training and competence.

The capacity of a road junction is dependent upon many of the features that govern link capacity with the addition of the junction type, control method and vehicle turning proportions.

The expression 'level of service' when applied to a highway refers to the *Highway Capacity Manual*[1] approach which defines a range of levels from the lowest which occurs during heavy congestion, to the highest, where vehicles can travel safely at their maximum legal speed (Table 5.1).

5.2 Effect of width on capacity

The capacity of a traffic lane is, within limits, proportional to its width. Clearly there is a lower limit to the width of a lane below which it is operationally impractical to run vehicles. Below a

Table 5.1 Maximum service flow rates

Design speed (mph)	Level of service	Density (PCU/mi/ln)	Speed (mph)	Maximum service flow rate
70	A	≤ 12	≥ 60	700
	B	≤ 20	≥ 57	1100
	C	≤ 30	≥ 54	1550
	D	≤ 42	≥ 46	1850
	E	≤ 67	≥ 30	2000
	F	> 67	< 30	unstable
60	B	≤ 20	≥ 50	1000
	C	≤ 30	≥ 47	1400
	D	≤ 42	≥ 42	1700
	E	≤ 67	≥ 30	2000
	F	> 67	< 30	unstable
50	C	≤ 30	≥ 43	1300
	D	≤ 42	≥ 40	1600
	E	≤ 67	≥ 28	1900
	F	> 67	< 28	unstable

lane width of about 2.0 metres, capacity deteriorates rapidly. As lane widths approach the point where two narrow lanes can be marked or vehicles tend to form up in two lanes there is a rapid increase in capacity. In urban areas and at road junctions traffic will tend to form up in two lanes when the lane width exceeds 5.0 metres. The effect of lane width and saturation flows at traffic signals was demonstrated by Kimber et al.[2] Figure 5.1 shows the simplified relationship between lane widths and saturation flow.

5.3   Effect of gradient

A steep uphill gradient can significantly affect the acceleration rate of all vehicles when pulling away from stationary at road junctions. Heavy vehicle speed also deteriorates on a combination of gradient and length of gradient. Additional crawler lanes are provided on long steep gradients on motorways and other heavily trafficked roads to maintain speeds and capacity. Webster & Cobbe[3] recognised this factor in their work on traffic signal saturation flows and further work in TRRL RR67.

5.4   Effect of alignment

A tightly curving alignment in rural areas can cause a reduction in free flow speeds. On existing roads tight curves are often accompanied by poor sight lines and forward visibility; that prevents slow moving vehicles from being overtaken and reduces overall capacity. Bunching of vehicles, with reduced headways, can cause excessive delays at side-road junctions.

In urban areas, curvature has been used to contain speeds in new residential areas and, more recently, artificial curves or horizontal deflection have been used to reduce vehicle speeds as part of traffic calming schemes. This is discussed in Chapter 11.

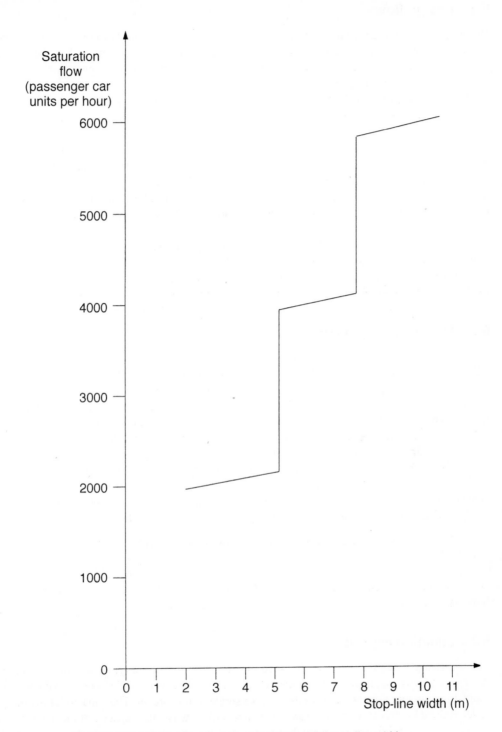

Fig. 5.1 Relationship between saturation flow and full stop-line width.

5.5   Design flows

Design traffic flow is an arbitrary value chosen by the highway authority to reflect the highway capacity and prevailing local conditions and includes a number of parameters such as the acceptable journey speed, free flow conditions, acceptable delay, environmental impact, minimum levels of safety for vehicular and nonvehicular traffic.

The highway authority (HA) will consider the existing flows on a route, expected traffic growth, traffic generation from changes in land use and desirable modal split. Where an existing highway is to be improved, the potential to improve the route and its junctions will affect the decisions on design flow. The design flows generally apply to a defined design year and usually refer to a minimum traffic handling capacity for the improved route. Increasingly HAs are considering maximum capacity and insisting that sustainable methods of transport are provided to limit the numbers of vehicle trips. Restraint on the numbers of car parking spaces at a development and subsidised public transport services are now an integral part of the design flow selection process.

A single minimum figure for design flow is now rarely sufficient to define the design parameters for a highway scheme. Usually a designer will be expected to provide space and capacity for alternatives to private modes of transport. Department of the Environment Guidance Notes PPG6[4] and PPG13[5] emphasise the move towards sustainable development and transport.

5.6   Flow–capacity relationships

A measure of the performance of a highway or junction is the ratio of demand flow to capacity (RFC) or traffic intensity. As the RFC approaches 1.0 the level of congestion and queuing will increase. There are two basic opinions on the effects of traffic intensity on queue lengths. The steady state theory suggests that as traffic intensity approaches 1.0 queue length will approach infinity. The deterministic theory assumes that no queuing occurs until the RFC exceeds 1.0. In practice it can be easily observed that queuing starts to occur well before an RFC of 1.0 is reached and it is equally clear that queue length does not approach infinity at this point. The Transport Research Laboratory (formerly Transport and Road Research Laboratory), in its junction modelling computer programs ARCADY[6] and PICADY,[7] has used a transformed curve for time-dependent queuing theory (Figure 5.2). This curve approximates to observable queues.

Figure 5.1 shows the theoretical relationship between flow and traffic density D:

$$D = \frac{\text{average number of vehicles in a length of highway } (L)}{L}$$

When the density is zero, flow is also zero; when the density increases to a maximum there is no flow. Maximum flow occurs at some point between these values.

5.7   Junction capacity

The traffic movements at uncontrolled road junctions fall into three types of conflict: merges, diverges and crossing movements. At merges, two traffic streams travelling in approximately the same direction join together and combine into a single traffic stream. The capacity of the merge is determined by the capacity of the two upstream carriageways, the capacity of the downstream carriageway, the traffic intensity and the relative speeds each traffic stream. At diverges, a single traffic stream separates into two traffic streams. Similarly, the capacity of the diverge is governed by the capacity of the upstream and down stream carriageways or lanes. Merges and diverges are

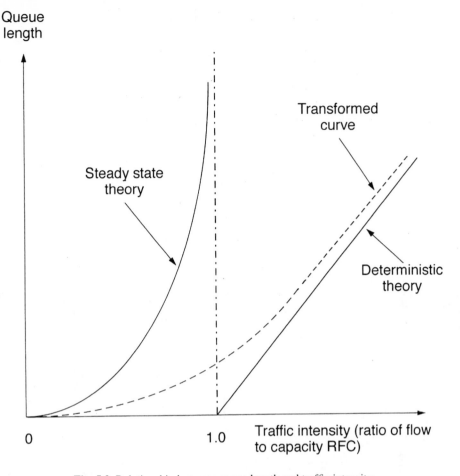

Fig. 5.2 Relationship between queue length and traffic intensity.

frequently accompanied by priority road markings and signs, such as those on the entries and exits to motorways or expressways. Occasionally there are more than two entries to a merge or more than two exits at a diverge but this is unusual and can be hazardous and confusing for drivers.

Crossing conflicts occur where two traffic streams cross each other at an acute angle. Usually one of the traffic streams has priority over the other. The capacity of the nonpriority stream is governed by the speed and intensity of the priority traffic stream. The driver on the nonpriority arms of a junction has to observe a suitable gap in the priority stream or streams before entering or crossing the priority stream.

5.8   Merges and diverges

Although merges occur within priority junctions and small roundabouts, the 'give-way' control generally requires the minor stream to slow or stop, and wait for a suitable gap, prior to entering the priority stream. On more major roads the merge is provided with an acceleration lane which enables drivers entering the priority stream to synchronise their speeds and make use of a smaller gap. If the diverge is provided with a suitable length of deceleration lane the 'through' priority

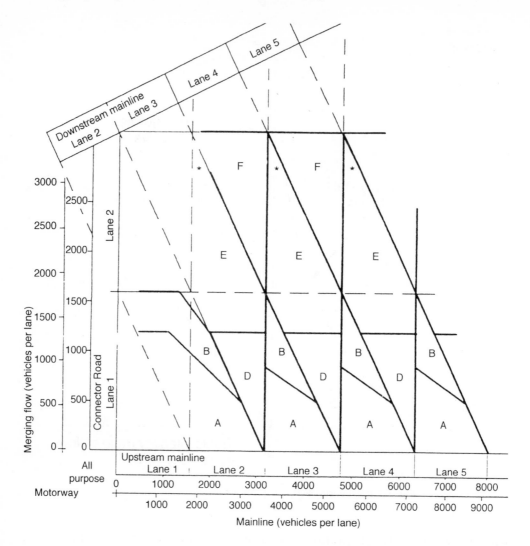

Fig. 5.3 Merging diagram for flow regions.

stream is largely unaffected. Merges and diverges are usually marked with a broken white line to indicate the edge of the priority stream. Figures 5.3 and 5.4, from TD 22/92,[8] show merge and diverge diagrams which are used with Figures 5.5 and 5.6, respectively, from TD 22/92.

5.9   Weaving sections

Weaving sections occur when a merge is followed by a diverge. Weaving sections are common on old style roundabouts and gyratories and between motorway interchanges. Weaving also occurs in urban streets. The capacity of a weaving section is governed by the length and width of the weaving section, the geometry of the upstream and downstream merges and diverges, and the proportion of weaving and nonweaving traffic in the section (see Figure 5.7 from TD 22/92[8]).

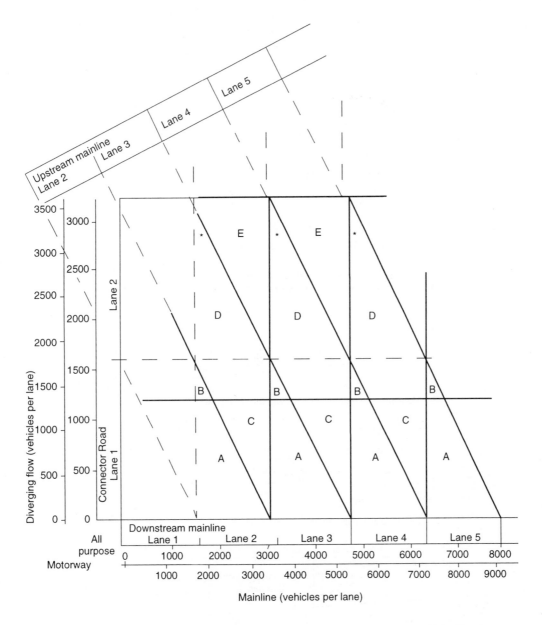

* Consider extended Auxiliary Lane

Fig. 5.4 Diverging diagram for flow regions.

Merge with no lane gain

    A   Taper merge

Nose
(2) & (3)

Entry taper
(1)

    B   Parallel merge

Nose
(2) & (3)

Auxiliary lane
(4)

Auxiliary lane taper
(4)

    C   Ghost island merge

Nose
(2) & (3)

Entry taper
(slip road right hand lane)
(1)
lane width 3.7 m

Ghost island taper
(6)
Ghost island width
2 m min. at
widest point

Entry taper
(1)

(Layout C is only used where design flows on mainline are light, there are 3 lanes or more on mainline and merging flow exceeds over one lane capacity)

Merge with lane gain

    D   Mainline gain

Nose
(2) & (3)

    E   Mainline lane gain at ghost island merge layout

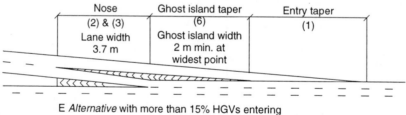

Nose
(2) & (3)
Lane width
3.7 m

Ghost island taper
(6)
Ghost island width
2 m min. at
widest point

Entry taper
(1)

E *Alternative* with more than 15% HGVs entering

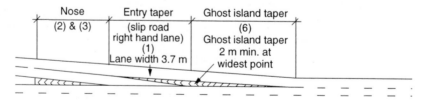

Nose
(2) & (3)

Entry taper
(slip road right hand lane)
(1)
Lane width 3.7 m

Ghost island taper
(6)
Ghost island taper
2 m min. at
widest point

(Layout F is the same as layout E but with both lanes added – the ghost island may be omitted)

N.B.   Figures in brackets refer to colums in Table 4/4 of TD 22/92

Fig. 5.5 Merge with and without lane gain.

Diverge with no lane drop

    A   Taper diverge

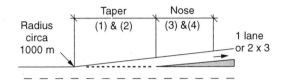

    B   Parallel diverge

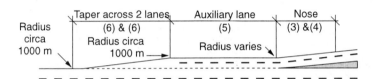

Diverge with lane drop

    C   Mainline lane drop at taper diverge

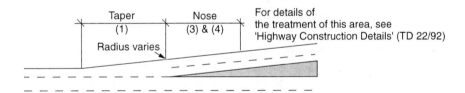

    D   Mainline lane drop at parallel diverge

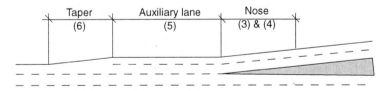

    (Layout E is layout D with both lanes off)

N.B. Figures in brackets refer to columns in Table 4/5 of TD 22/92

Fig. 5.6 Diverge with and without lane drop.

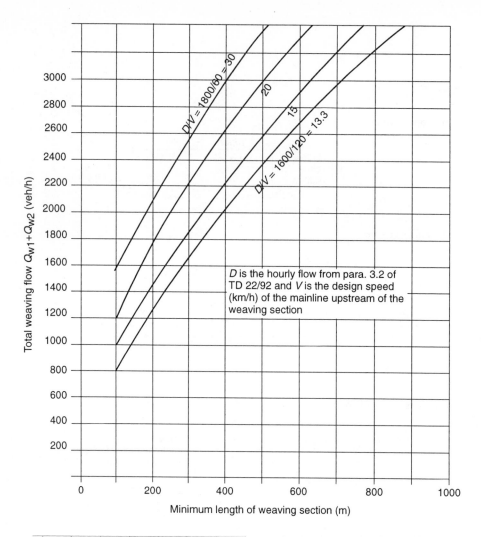

D is the hourly flow from para. 3.2 of TD 22/92 and V is the design speed (km/h) of the mainline upstream of the weaving section

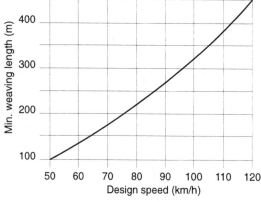

To determne the minimum length of weaving section (L_{min}) or insertion within the formula of paragraph 2.26 of TD 22/92:

1. For known total weaving flow and chosen D/V value, read off the minimum length of weaving section from the graph above.

2. Check the minimum weaving length allowable for chosen design speed from the graph on the length.

3. Select the greater of the two lengths.

Fig. 5.7 Total weaving flow versus minimum length of weaving section.

References

1. Transportation Research Board (1985) *Highway Capacity Manual*, Special Report 209, TRB, Washington, DC.
2. Kimber, RM, McDonald, M and Hounsell, NB (1986) *The prediction of saturation flows for road junctions controlled by traffic signals*, TRRL Research Report 67, TRRL, Crowthorne.
3. Webster, FV and Cobbe, FM (1966) *Traffic Signals*, Road Research Technical Paper 56, HMSO, London.
4. Departments of the Environment and Transport (1994) *Planning Policy Guidance – Town centres and retail developments*, (PPG 6), HMSO, London.
5. Departments of the Environment and Transport (1994) *Planning Policy Guidance – Transport*, (PPG 13), HMSO, London.
6. Department of Transport (1985) *ARCADY2: An enhanced program to model capacities queues and delays at roundabouts*, TRRL Research Report 35, TRRL, Crowthorne.
7. Department of Transport (1985) *PICADY2: An enhanced program to model capacities queues and delays at major/minor priority junctions*, TRRL Research Report 36, TRRL, Crowthorne.
8. Department of Transport (1992) *Layout of Grade Separated Junctions*, TA 48/92 and TD 22/92, HMSO, London.

6

Traffic Management and Control

6.1   Objectives

Traffic management arose from the need to maximise the capacity of existing highway networks within finite budgets and, therefore, with a minimum of new construction. Methods, which were often seen as a quick fix, required innovative solutions and new technical developments. Many of the techniques devised affected traditional highway engineering and launched imaginative and cost effective junction designs. Introduction of signal-controlled pedestrian crossings not only improved the safety of pedestrians on busy roads but improved the traffic capacity of roads by not allowing pedestrians to dominate the crossing point.

More recently the emphasis has moved away from simple capacity improvements to accident reduction, demand restraint, public transport priority, environmental improvement and restoring the ability to move around safely and freely on foot and by pedal cycle.

6.2   Demand management

There has been a significant shift in attitudes away from supporting unrestricted growth in highway capacity. The potential destruction of towns and cities and the environmental damage to rural areas is not acceptable to a large proportion of the population. Traffic management has, largely, maximised the capacity of the highway network, yet demand and congestion continues to increase.

Highway authorities accept that they do not have a mandate to provide funds for large amounts of new construction. It is clear that, for the foreseeable future, resources will not be available to provide for unrestricted growth in private vehicular traffic. Traffic engineering alone cannot provide sufficient highway capacity even with limited amounts of new construction.

As traffic demand and congestion increased, drivers found alternative routes, often though residential areas. Road safety was compromised as drivers travelled at high speed to maximise the benefits of diverting from their normal route. Pressure from residents, in these areas, led to the introduction of area wide environmental traffic management schemes (ETMS) during the 1970s an 1980s.

ETMSs attempted to deny these rat-runs to queue-jumping traffic and to specific classes of vehicle such as wide or heavy vehicles. Many ETMSs were spectacularly effective and used such techniques as point road closures, physical width restrictions, one-way plugs, one-way streets and banned turns. These measures, which were designed to be restrictive for unwelcome traffic, often caused great inconvenience to residents, emergency vehicles and service vehicles.

Frequently, residents were prepared to tolerate severe inconvenience in order that a safe and tranquil environment could be restored.

In recent years the emphasis has moved towards reducing vehicle speeds and accidents using traffic calming and reallocation of road space. ETMS techniques are still widely used often in combination with traffic calming methods. Often the broad aims of ETMSs can be achieved by traffic calming alone. Traffic calming is discussed in detail in Chapter 11.

In town centres whole carriageways have been closed to general traffic and dedicated to pedestrians, cars driven by people with disabilities, buses and service vehicles. Many town centre schemes were preceded by hostile objections from shopkeepers concerned about loss of passing trade. However, most town centre schemes have produced a much improved trading environment. Outside shopping hours many centres appear lifeless and, in some instances, general traffic has been allowed back into the centre during those hours.

Figure 6.1 shows a typical daily flow profile for a radial route into a large town or city. It can be seen that peak periods are relatively short and that for long periods of the day traffic flows are well below the road's capacity.

As delays increase, drivers realise that there is spare capacity at other times of the day and change their working hours to arrive at and depart from their places of work before and after the peaks. The result of this is that peaks last longer, an effect known as peak spreading. Ultimately the peaks spread until the morning and afternoon peaks meet. The daily flow profile in Figure 6.2 shows the majority of daytime capacity has been utilised. This effect is observable in parts of Central London and other large cities where heavy congestion occurs throughout the working day. Sunday trading has produced dramatic increases in traffic flows on what was a lightly trafficked day. The only remaining part of the day where capacity is available is during the night, even this is now under attack as some large superstores are staying open throughout the 24 hour day.

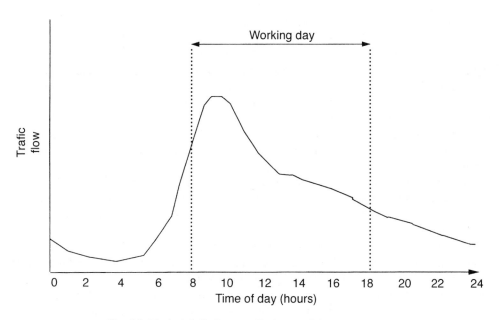

Fig. 6.1 Typical daily flow profile for a radial route into a city.

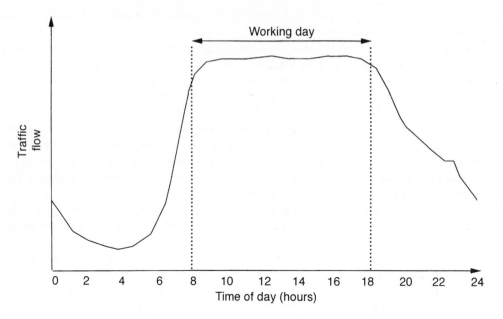

Fig. 6.2 Daily flow profile showing capacity fully utilised during the working day.

In many cities overall journey speeds are restricted to around 10 mph and have a severe effect upon the economic health of the centre, the physical health of the population and essential traffic such as the emergency services and public transport are severely impeded.

It is clear that unrestricted use of private cars cannot continue. Some limitation on the freedom of use is inevitable if towns and cities are not to become polluted wastelands clogged with slow-moving traffic for most of the day.

Removal of the need to travel by private car and removal of demand by means of parking and junction control with parallel incentives to use public transport offer a solution.

One of the most effective ways to restrict private car traffic to a city centre is on-street and off-street parking control. Parking control is discussed in Chapter 9. Restriction on numbers together with punitive charges forces drivers to consider alternative modes for their journeys. Control of private nonresidential (PNR) parking must accompany control of public car parks and on-street parking or many of the benefits will be lost.

The town planning and development control process must keep pace with parking policy, and there might be a need for additional legislation to ensure that neighbouring authorities produce compatible policies.

Reduction in highway capacity by transferral of road space to more efficient use by higher-capacity vehicles such as buses and trams is doubly effective. Bus lanes reduce journey times and provide a clear demonstration to car drivers of the efficiency of public transport. If a significant modal shift occurs bus journey times reduce further.

Reduction in highway capacity must be applied over a whole network, otherwise untreated and therefore unprotected routes will be used by drivers as alternative routes. This effect is similar to the rat-running problems encountered in the early days of ETMS. The wider area effects of such measures can be tested using traffic models such as TRIPS, SATURN and QVIEW. Sometimes, where the network being considered is relatively compact, carefully monitored experiments are more productive. In large conurbations and major cities diversionary effects can occur at a great distance from the site under consideration. Drivers discouraged from using a radial route may divert to another radial, using a suborbital route (as on the M25 in London). In most cases the set-tling down or stabilising of traffic flows can take many weeks. Local congestion caused by an experimental restraint measure can create intense local pressure for its removal, often before the new traffic patterns have been established.

One method of demand management that has received considerable interest and study is con-gestion charging. This is where vehicles are surcharged for their use of road space depending on the prevailing levels of congestion. New technology in the form of smart cards and vehicle iden-tification are needed to ensure that the system is practical and fair. Measures to accommodate for-eign or nonlocal vehicles are also needed. Video image analysis has reached a level of sophistication that can be used for this purpose.

Detailed congestion-charging experiments and studies have been carried out in London[1] and Cambridge, but no authority has yet attempted to introduce a citywide system in the UK. The studies have demonstrated that practical systems could be introduced within the next 10 years.

6.3   Engineering measures

The traffic engineer has a vast array of measures which can be applied to achieve his objectives. These objectives include:

- capacity enhancements
- accident reduction
- environmental protection and enhancement
- servicing of premises and providing access
- providing assistance to pedestrians and cyclists
- assisting bus or tram operators
- providing facilities for persons with disabilities
- regulating on-street and off-street parking.

The majority of capacity problems occur at road junctions. In urban areas road junctions are important focal points for pedestrian and cycling activity and are often the site of public transport interchanges. Due to the various conflicting demands it is not surprising that two-thirds of urban traffic accidents occur at road junctions. Selection of an appropriate junction design for a partic-ular site can be very difficult. Some designs, such as roundabouts, can significantly reduce the severity of vehicle–vehicle accidents but can prove hazardous for cyclists. In some instances installation of traffic signals with full pedestrian and cycle facilities and bus priority measures might also reduce the overall traffic-handling capacity.

Careful allocation of road space to separate traffic streams into designated traffic lanes can reduce confusion and limit accidents. Designation of traffic lanes might include special vehicle lanes, such as cycle and bus lanes and dedicated left- or right-turn lanes.

Introduction of banned turns and one-way streets can reduce potential conflicts and accident potential. These measures can be used to implement protected pedestrian or cycle crossings and simplify junction layouts generally. Great care must be exercised when one-way street schemes are being considered as they can result in speeding by drivers who are confident that they will not be opposed by other vehicles.

Point road closures are used to simplify junction and highway layouts and eliminate turning conflicts. The resulting continuous footway can also improve pedestrian safety and provide space for bus stops, cycle racks, pedestrian crossings, and hard and soft landscaping.

Closure of long sections of road to general traffic can produce pedestrianized shopping streets.

Such schemes can be very complex to design and introduce, because facilities for buses, emergency services, residents/proprietors and service vehicles must be considered.

Carriageway narrowing can be used to limit capacity or vehicle speeds and reduce parking and pedestrian crossing distances. Carriageway narrowing is discussed in Chapter 11.

The key to all successful traffic engineering schemes is that the visual cues provided by the road must give a clear indication to users of who has priority.

6.4   Junction types

There are many varying detailed junction types, but they can be broken down into five basic types:

- uncontrolled nonpriority junctions
- priority junctions
- roundabouts
- traffic signals
- grade separation.

Each junction type can accommodate different levels of major and minor road flow. Choice of a particular junction type depends upon the flow levels and the space available for its construction or reconstruction, as shown in Figures 6.3 to 6.7.

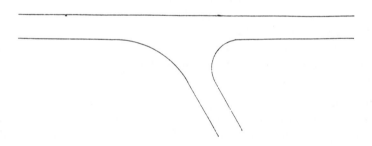

Fig. 6.3 Uncontrolled non-priority junction.

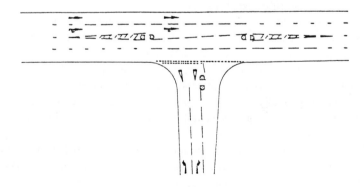

Fig. 6.4 Priority junction.

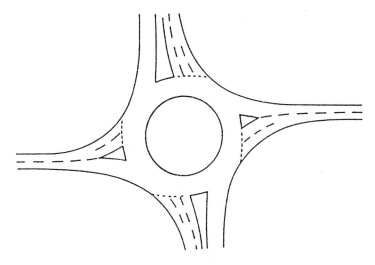

Fig. 6.5 Roundabout.

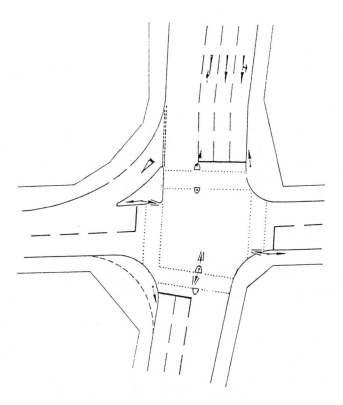

Fig. 6.6 Traffic signals.

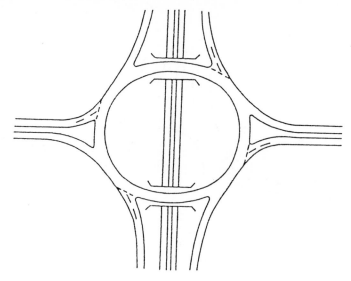

Fig. 6.7 Grade separation.

6.4.1   UNCONTROLLED NONPRIORITY JUNCTIONS

At the lowest flow levels, such as those experienced in small residential areas, nonpriority junctions are suitable. However, even at very low flow levels, a lack of priority can lead to confusion and accidents. For modest costs, simple priority road markings can remove uncertainty.

6.4.2   PRIORITY JUNCTIONS

These vary from very simple T junctions, with give-way markings, catering for low flows, to highly complex junctions on single- or dual-carriageway roads, with turning movements separated by channelisation islands, ghost islands and auxiliary priority markings and signs within the overall junction area. If sufficient space is available, very heavy vehicular flows and high speeds can be handled safely.

6.4.3   ROUNDABOUTS

Roundabouts can be considered as a special type of priority junction. Vehicles give way to off-side traffic and circulate around a central island (in a clockwise direction in the UK). Roundabouts can be very large – up to 200 metres or more across and as small as 13 metres across.

The smallest roundabouts are called mini-roundabouts and the central island is reduced to a raised, drive-over dome or a simple flat circle, usually painted white and accompanied by a series of circulating arrows marked around the centre.

Though the very small and large roundabouts both have offside priority markings they function quite differently. The circulating movement is almost absent in the smaller junctions, and straight-across movements become crossing conflicts. Roundabouts were a particularly British device but are now in common use in most countries; France in particular has introduced them widely now that they have resolved their nearside priority problems.

6.4.4   TRAFFIC SIGNALS

Traffic signals were originally used when there was insufficient land available for enlarging a junction. Increasing sophistication in the control systems has resulted in their widespread use in heavily trafficked urban areas. Positive control is now applied to different road users such as pedestrians, cyclists and buses.

Traffic signals can operate on fixed time plans where the time allowed for each arm is preset, using historical traffic data. Vehicle actuation uses detectors to measure traffic demand on the approaches to the junction and vary the amount of green time allotted to each arm. Green time can be varied between a minimum and maximum time that is preset in the controller. Under conditions of low flow, detection of a vehicle can initiate a green stage.

There are many examples where signals have been installed on roundabouts with severe congestion problems. For many years traffic signals have been linked over wide areas to minimise delay and maximise network capacity. Linking can be simple synchronisation of successive sets of signals to produce a green wave for vehicles along a route. Fixed time plan urban traffic control and adaptive systems such as SCOOT[2] have been used to maximise network capacity for many years.

6.4.5   GRADE SEPARATION

Where a particular movement through a junction has a very high flow it can be separated vertically from the other turning movements. The simplest grade separation, at a crossroads, is the diamond interchange. The minor road passes under or over the major road and slip roads connect to the major road from an at-grade junction on the minor road. Depending on the flow levels, the minor at-grade junction could be controlled by a priority junction, a roundabout or roundabouts or traffic signals.

Where two or more major highways cross, the connections are often made by connecting the slip roads together and avoiding at-grade conflicts altogether. Grade separation is used on all motorways in the UK and on other major highways. Some examples of grade-separated junctions are shown in Figure 6.8.

6.5   Road markings

It is not possible to overestimate the importance of road markings as part of the road system. In a few instances road markings merely emphasise the layout of the highway and guide road users to a safe course of action. In many cases the whole of the success of a scheme relies upon the visual messages emanating from the road markings (miniroundabouts are often implemented by road markings alone supplemented by a few traffic signs).

Carriageway markings in the UK must be placed in accordance with the *Traffic Sign Regulations and General Directions 1994*[3] (*TSRGD*). Further advice is given in the *Traffic Signs Manual*[4].

Road markings must be designed into a scheme at the earliest stage; they cannot be added later when all the other aspects of a scheme have been agreed. Usually the locations of edge of carriageway, lane lines, ghost islands and priority markings are as critical as the location of the kerb lines, traffic islands and other highway features. Road markings not only guide road users but provide evidence of traffic regulations, such as waiting and loading restrictions, pedestrian crossings, box junctions, keep clear markings and level crossings.

In rural areas the double-white-line system uses double longitudinal lines. Solid white lines are used to prevent dangerous crossings of the carriageway centre line. In urban areas solid lines delineate bus and cycle lanes and stop lines. Yellow lines (and in London red lines) on the carriageway parallel to the kerb indicate waiting restrictions; loading restrictions are indicated on the kerb itself. Waiting and loading restrictions markings on kerbs are common in the USA and many other countries.

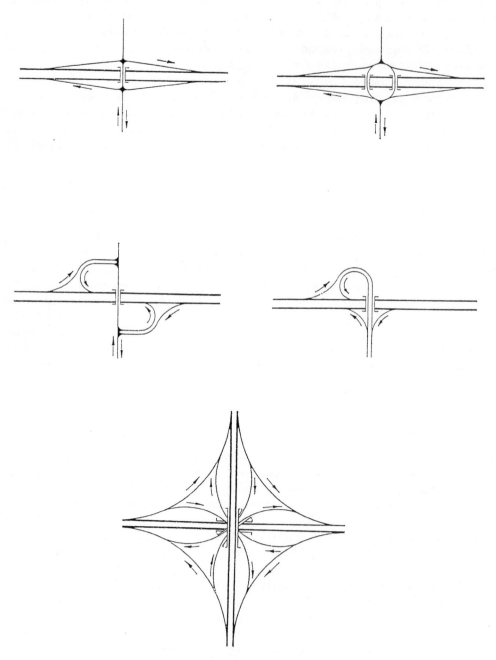

Fig. 6.8 Grade-separated junctions.

6.6 Traffic signs

Traffic signs fall into four categories:

- warning signs
- regulatory signs
- directional informatory signs
- other informatory signs.

Warning signs provide information to road users about hazards such as junctions, changes of direction, carriageway width, gradient, low, opening or humped back bridges, roadworks etc.

Regulatory signs provide a message that must be obeyed, such as stop, give way, banned turns, compulsory turns, no entry, one-way streets, prohibited vehicle types, weight and width restrictions, waiting and loading restrictions, speed restriction etc.

Directional informatory signs provide information about routeing and important places of interest, such as railway stations, airports etc.

Other informatory signs provide information about footway and other parking schemes, heritage sites, census points etc. Traffic signs are often installed in association with road markings (give-way triangle, waiting and loading plates).

Figures 6.9–6.12 show typical UK road signs. Details about location of signs and their design are given in the *TSRGD 1994* and *Traffic Signs Manual*. It is important that the advice is followed closely, to ensure compatibility throughout the UK.

In certain circumstances special signs or road markings might be needed and authorisation can be sought from The Secretary of State for Transport in England, The Secretary of State for Scotland or The Secretary of State for Wales. All signs placed on the public highway must be in accordance with the regulations (or specially authorised by the Secretaries of State).

In recent years variable message signs (VMS) have become more common. These signs are controlled by car park entry systems, motorway control centres, urban traffic control centres and, more recently, by speed cameras, CCTV and video imaging systems.

There are two basic types of VMS in common use in the UK: rotating boards and dot-matrix signs. Rotating board signs can have double sided or triangular prism signs. These signs are clear and easy to read as they appear very similar to static road signs; their main disadvantage is that

507.1
Staggered junction ahead

510
Roundabout ahead

Fig. 6.9 Warning signs.

611.1

Roundabout

612

No right turn for vehicular traffic

Fig. 6.10 Regulatory signs.

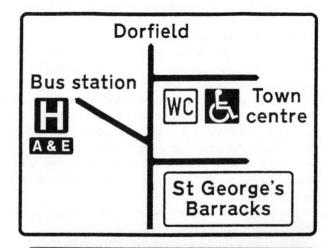

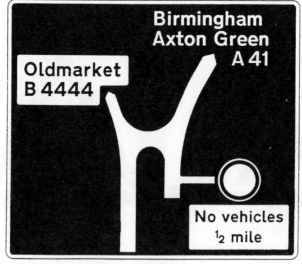

Fig. 6.11 Directional informatory signs.

Fig. 6.12 Other informatory signs.

they are restricted to two or three fixed messages. Dot-matrix signs use fibre optics to display a wide variety of messages and are thus very flexible in operation. However the operator must be careful to use easily read unambiguous messages. Modern signs are reasonably easy to read except in conditions of direct, bright sunlight. Recently there have been moves to use sponsored dot-matrix signs for commercial advertising when they are not needed for traffic purposes. This use might have road safety implications.

References

1. Gilliam, C and Richards, M (1996) The London Congestion Charging Research Programme, a series of papers published in *Traffic Engineering & Control*: **37**(2), 66–71; **37**(3), 178–179, 181–183; **37**(4), 277–282; **37**(5), 334–339; **37**(6), 403–409.
2. Department of Transport (1981) *SCOOT: a traffic responsive method of coordinating signals*, TRRL Laboratory Report 1014, TRRL, Crowthorne.
3. UK Government (1994) *Traffic Signs Regulations and General Directions 1994* (Statutory Instrument 1994 No 1519), HMSO, London.
4. UK Government (1974–) *The Traffic Signs Manual* (separate Chapters 1 to 14 not all available) HMSO, London.

7

Highway Layout and Intersection Design

7.1 Highway link design standards

It is unlikely that the traffic engineer alone will be required to design major highways. Highway design is a separate, albeit a closely related, discipline. At most the traffic engineer will provide preliminary layouts of access roads and intersections for developments and, therefore, a good understanding of basic highway design techniques and standards is needed.

7.1.1 DESIGN SPEED

Selection of an appropriate design speed may be considered the starting point for any scheme. The DoT Technical Standard TD 9/81[1] and Advice Note TA 43/84[2] outline the methods for selecting the link design speed. In practice the design speed for a particular route might be contained within a policy decision by the highway authority. The engineer should consult the HA closely and, if no advice is forthcoming, suggest a method for determining a suitable design speed. In this case the design speed should be selected by observation of the actual behaviour of the vehicles on the road in question.

Vehicle speeds are affected by many factors including speed limit, horizontal and vertical alignment, visibility, highway cross-section, adjacent land use, spacing of junctions, accesses, pedestrian crossings and maintenance standards. The general condition and design of vehicles and driver ability, which change over time, can have a significant effect on vehicle speeds. It is usual to use an 85th percentile speed as the design speed (the speed below which 85% of drivers travel). The 85th percentile speed is determined from speed surveys using a radar speed meter or automatic traffic counter equipped to measure speeds.

In urban areas, where the speed limit is less than the national speed limit, the design speed is more likely to be based upon a policy decision. Frequently the traffic engineer is asked to reduce the speeds of vehicles to an acceptable level. In these circumstances it would be wholly inappropriate to use the 85th percentile speed as the design speed because this is often well above the existing or proposed speed limit. The engineer may be instructed to introduce measures that reduce or control 85%, or an even higher percentage, of vehicle speeds to the speed limit.

7.1.2 HORIZONTAL CURVATURE

The radius of curvature that a vehicle can travel round at the design speed depends upon the crossfall or cant of the carriageway and adhesion between the tyres and the road surface. The

Department of Transport provides a table of standards, reproduced as Table 7.1, for standard design speeds for different crossfalls or superelevation.[1] Figure 7.1 shows the forces acting upon a vehicle travelling around a curve.

The centrifugal force acting on the vehicle is

$$\text{centrifugal force} = \frac{Mv^2}{R}$$

where

M = mass of the vehicle
v = velocity in metres/second
R = radius of curvature

The force in the vertical direction is

$$\text{vertical force} = Mg$$

where g is the acceleration due to gravity. If the forces acting parallel to the road surface are to equal each other, then

$$\frac{Mv^2}{R} \cos \alpha = Mg \sin \alpha$$

Table 7.1 Recommended geometric design standards

Design parameter[a]	Design speed (km/h)					
	120	100	85	70	60	50
A: Stopping sight distance (m)						
A1 Desirable minimum	295	215	160	120	90	70
A2 Absolute minimum	215	160	120	90	70	50
B: Horizontal radii (m)						
B1 Minimum radius[b] without elimination of adverse camber and transitions	2880	2040	1440	1020	720	510
B2 Minimum radius[b] with superelevation of 2.5%	2040	1440	1020	720	510	360
B3 Minimum radius[b] with superelevation of 3.5%	1440	1020	720	510	360	255
B4 Desirable minimum radius[b] with superelevation of 5%	1020	720	510	360	255	180
B5 Absolute minimum radius[b] with superelevation of 7%	720	510	360	255	180	127
B6 Limiting radius[b] with superelevation of 7% at sites of special difficulty (category B design speeds only)	510	360	255	180	127	90
C: Vertical curvature (m)						
C1 FOSD overtaking crest K value	[b]	400	285	200	142	100
C2 Desirable minimum[b] crest K value	182	100	55	30	17	10
C3 Absolute minimum[b] crest K value	100	55	30	17	10	6.5
C4 Absolute minimum sag K value	37	26	20	20	13	9
D: Overtaking sight distance (m)						
D1 Full overtaking sight distance	[b]	580	490	410	345	290

[a] FOSD = Full overtaking sight distance; K value is used to calculate the lengths of vertical curves.
[b] Not recommended for use in the design of single carriageways.

therefore,

$$\tan \alpha = \frac{v^2}{gR}$$

where

α = superelevation (cant or crossfall)

In practice these forces are rarely in perfect balance and the differences are accommodated in the friction between the tyres and the road surface.

Generally, the desirable minimum radius should be used for new construction. If there are valid cost or environmental reasons, a relaxation of standards to the absolute minimum radius can be permitted. The limiting radius can only be used under certain circumstances at difficult sites. Departures from standards are not recommended except in urban areas where full standards cannot be achieved. In these circumstances it might be possible to use traffic engineering and traffic calming techniques to mitigate the effects of substandard alignment (Chapter 11).

7.1.3   TRANSITION CURVES

A transition curve varies in radius from the straight line to the horizontal curvature of the road. Various curve forms have been used such as a true spiral, the cubic parabola and the lemniscate. The rate of increase of centripetal or radial acceleration should normally be limited to 0.3 m/s^3

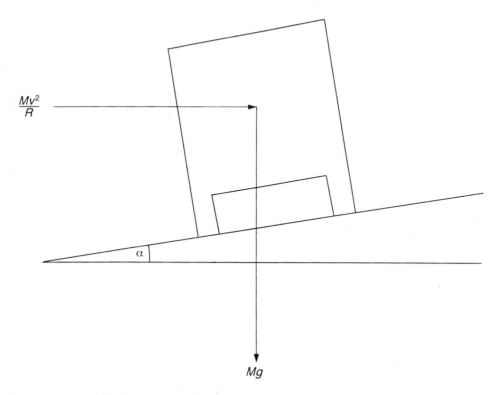

Fig. 7.1 Forces acting upon a vehicle travelling around a curve.

but can be increased to 0.6 m/s³ at constricted sites. Where superelevation is to be introduced or adverse camber eliminated, it should be done over the length of the transition curve. The transition curve starts approximately half its length before the normal circular curve tangent point and shifts the circular curve inwards towards its origin (Figure 7.2).

Superelevation improves safety and comfort within the vehicle and allows use of smaller radii. In urban areas superelevation should not, and often cannot, be applied slavishly. Road levels are usually constrained by frontage development and road junctions. In urban areas it is often more appropriate to provide a camber across the cross-section of the road that is sufficient for drainage purposes only.

Within road junctions, superelevation should be avoided where possible and should always be avoided at roundabouts where it encourages excessive speeds and can seriously affect visibility and understanding of the layout by drivers (Section 7.3).

7.1.4 VERTICAL ALIGNMENT

Vertical alignment consists of a series of straight gradients and connecting curves. The main constraints are comfort and visibility over crest curves. Climbing vehicles are slowed by steep gradients which should be avoided wherever possible. If long steep gradients cannot be avoided,

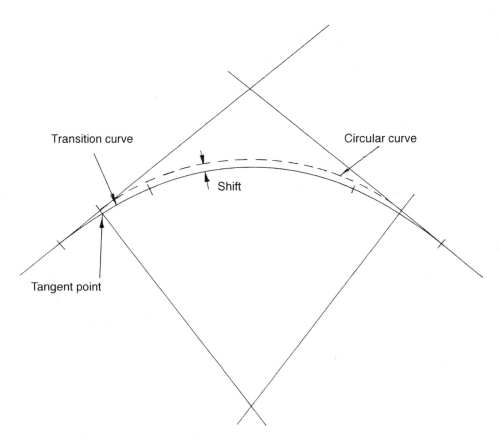

Fig. 7.2 Transition curve.

consideration should be given to the provision of crawler lanes for heavy vehicles. In urban areas, gradient is often constrained by side roads and buildings. In many towns and cities very flat roads may force the introduction of artificial gradients or false channels simply to provide adequate surface run-off.

The length of vertical crest and sag curves are calculated using the design speed coefficient K obtained from Table 7.1[1] and the difference in the percentage gradients A:

length of vertical curve L (m) = KA

Note that a rising gradient from left to right is positive and a falling gradient is negative.

7.1.5   SIGHT DISTANCES

Table 7.1 gives two sight distances:

- stopping sight distance (SSD)
- full overtaking sight distance (FOSD).

Stopping sight distance is the distance required by drivers to stop their vehicles if they see an unexpected obstruction on the road surface. The SSD includes the 2 seconds driver reaction time and the braking distance on a wet road. The vertical SSD envelope is shown on Figure 7.3. The driver's eye height is between 1.05 m and 2.0 m above the road surface.

The horizontal SSD envelope is measured from the centre of the nearside lane across the bend. To obtain the envelope the SSD is measured at points around the curve from positions where the sight line lies within the carriageway. On very tight curves the need to acquire additional land (land take) for SSD can be considerable and sometimes the provision of a larger radius of curvature results in less land take.

On rural single carriageways it is sometimes necessary to provide FOSD. The details of its application are given in TD 9/81[1] It is unlikely that the majority of traffic engineers will need to design a scheme with FOSD, especially in urban areas.

In urban areas there is immense pressure to place street furniture on footways and verges which could compromise the SSD.

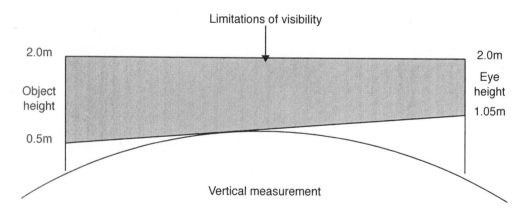

Fig. 7.3 Measurement of stopping sight distance.

7.1.6 INTERSECTIONS – GENERAL

In rural areas there is often space to provide generous layouts for new junctions. Land cost is relatively low and it is usually possible to adjust the overall scheme layout to achieve full design standards. There are, of course, constraints such as the difficult terrain found in mountainous or hilly areas that physically restrict the freedom of the engineer to optimise designs. There are other constraints which have increased in importance in recent years such as:

- areas of outstanding natural beauty (AONB)
- sites of special scientific interest (SSSI)
- listed buildings and conservation areas
- archaeological sites
- high-quality farm land.

Even with these constraints, it is important and usually possible to achieve close adherence to design standards. Design speeds in rural areas are usually higher than in urban areas and the consequences of accidents and collisions at higher speeds are more severe.

Within urban areas the physical constraints are more related to the built environment. Property values are higher and opportunities to appropriate land for highway improvements are fewer. Even within the highway boundary the presence of underground services can impose conditions upon the layout. With some small traffic schemes, the cost of protecting or rerouteing services can be 10 times (or more) than the cost of the highway and signing works.

With any urban scheme the most important consideration is the size, turning characteristics and space requirements of the design vehicles. In most instances this means the largest legally permitted road vehicles. In residential areas the predominate vehicle will be the private car, but large vehicles regularly penetrate these areas for refuse collection, gulley cleaning, maintenance, household deliveries, school buses and public transport. In town centres, department stores, supermarkets, public houses and fast food outlets receive regular deliveries by the largest heavy goods vehicles. Figure 7.4 shows typical large vehicle turning circles.

Occasionally, abnormal loads enter town and city centres. Major cities, such as London, have designated abnormal load routes that avoid weak or low bridges, difficult turning manoeuvres or overhead telephone or power lines. The engineer must be aware of these and ensure that provision is made, within new designs, for these eventualities. The additional cost of installing hardened overrun areas and demountable street furniture can sometimes be recovered from the operators of these outsized vehicles.

In very rare cases the design vehicle can be limited to less than the largest vehicles where physical width or weight limits exist. Where general traffic flows and HGV movements are low, the occasional large vehicle can be allowed to dominate the full width of a road for the short time it needs to complete its manoeuvres.

At urban road junctions it is often not possible to provide full standards for large vehicles. Where their numbers are small and speeds are low they can be allowed to dominate two or more traffic lanes at urban roundabouts and traffic signals. Traffic islands, kerb lines and street furniture must, of course, be placed outside the vehicle swept paths. Standard vehicle swept path plots are available for checking layouts but for more complex multiple turns the DoT computer program TRACK[3] is recommended.

Apart from abnormal load vehicles the longest vehicle in everyday use in the UK is the 18 m drawbar trailer combination. When designing junctions at confined sites, it is not enough to check that the layout is adequate for the largest vehicles. Swept path is dependent on a number of factors:

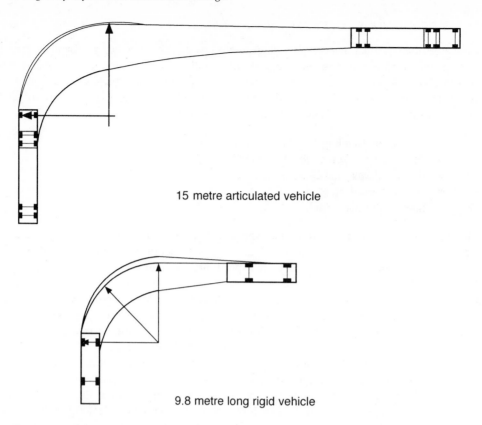

15 metre articulated vehicle

9.8 metre long rigid vehicle

Fig. 7.4 Heavy vehicle turning circles.

- rigid or articulated
- wheelbase (or tractor wheelbase)
- front and rear overhang
- width
- length of trailer.

In some instances an articulated vehicle has a smaller turning circle and smaller swept area than a long rigid vehicle. Vehicles with long front or rear overhangs, such as the 12 m European low floor bus can pose particular problems. If such a vehicle is stationary and its steering wheels are on full lock the rear will tend to move outwards as it starts to move. If the vehicle is close to the kerb the bodywork might strike a pedestrian.

7.1.7   SELECTION OF JUNCTION TYPE

Selection of junction type can be very simple and obvious in some cases, for example:

- two lightly trafficked residential roads – priority junction;
- the through carriageway of a motorway – grade separation;
- heavily trafficked urban crossroads with heavy pedestrian flows – traffic signals;
- suburban dual carriageways with substantial heavy goods traffic – conventional roundabout.

However, there are very many cases where the solution is anything but obvious and the engineer should resist making a decision until all the evidence has been examined and analysed.

Frequently, when an existing junction is to be upgraded to handle more or different types of vehicle, the existing form or control method must be considered. The form of the other junctions on the main route might determine the form of the new or improved junction; for example, it might be inappropriate to construct a miniroundabout on a route that has a series of linked

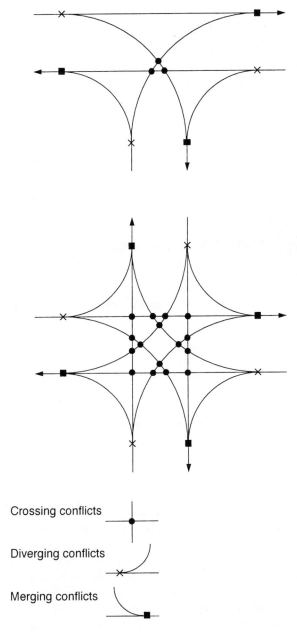

Fig. 7.5 Conflict points at a T junction and a crossroads.

signalled junctions and pelican crossings. Similarly a signalled junction on an otherwise free-flowing rural dual carriageway with generously designed priority junctions or roundabouts could prove to be hazardous.

All junction design is an iterative process which moves between space and capacity requirements. A good starting point is to look at all the required turning movements and their design flows. A conflict diagram as shown in Figure 7.5 is invaluable. The traffic flows can be superimposed on the conflict diagram to give an indication of the importance of each conflict (Figure 7.6). If possible the junction movements should be reduced to a series of three-arm T junctions. Some minor movements can be prohibited if there are suitable alternative routes for the prohibited traffic, (e.g. a crossroads becomes right/left staggered T junctions). Conflict diagrams should be sketched for all extremes of traffic flow, such as morning and evening peak periods or perhaps heavy weekend flows into a retail park or a sporting event. Because these flows might have completely different turning proportions, it is not sufficient to design for the highest throughput at the junction.

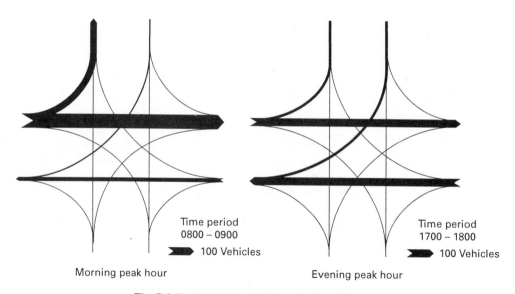

Fig. 7.6 Typical peak period junction flow diagrams.

7.2   Priority junctions

Priority junctions are the simplest and most common of intersections and range from single lane approach T junctions to high-capacity channelised layouts. The control at priority junctions depends upon give-way road markings and post mounted signs.

Crossroad layouts should be avoided wherever possible because they concentrate a large number of vehicle movements and, therefore, conflicts within the junction. On unlit rural roads at night, drivers on the side roads can be confused into thinking that they are on the major route, particularly when there is an approaching vehicle on the far side of the crossroads.

The DoT Advice Note TA 20/84[4] provides details on the layout of major/minor junctions. The TA 20/84 layouts will accommodate the largest vehicle swept paths and a simple check by the engineer will confirm that a design is satisfactory. In urban areas it might not be possible or desirable to achieve the full standards and more detailed checking of the layout will be necessary. If the generous turning radii are provided in areas of extreme parking pressure, motorists might be tempted to park within the junction area. This will compromise the layout and interfere with the swept paths. Waiting and loading restrictions or kerb buildouts can be used to limit this problem.

7.2.1   VISIBILITY

To enable drivers to safely cross or enter the major-road traffic streams they must be able to see and to judge their approach speeds and available gaps in the major-road traffic.

Visibility splays are provided from the side road to left and right. A triangular sight line envelope, measured along the centreline of the side road and along the major road kerbline must be maintained clear of obstruction above the driver's eye height of 1.05 metres. The visibility envelope is defined by its x and y distances (Figure 7.7).[4] The x distances should be a maximum of 15.0 m and a minimum of 9.0 m at junctions on major roads. The x distance can be reduced to 4.5 m where flows are low and an absolute minimum of 2.4 m. The y distance is dependent upon the major road design speed and its distance should not be compromised. The visibility splay joins the two points defined by x and y except on a curved alignment where it should be tangential to the carriageway edge (Figure 7.8).[4] Generally the visibility splay should be within the highway boundary to ensure that it can be maintained free of obstruction. Highway authorities can exercise powers under the Highways Act to maintain clear sight lines within the curtilage of private property where necessary.

Where the major road has been provided with a central reservation or ghost island for vehicles to wait, the visibility distance should also be provided.

At larger, more heavily trafficked junctions or where there are significant pedestrian movements, guide islands and pedestrian refuges can be provided. Channelisation islands help to

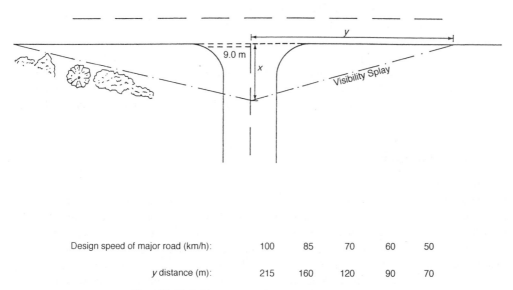

Design speed of major road (km/h):	100	85	70	60	50
y distance (m):	215	160	120	90	70

Fig. 7.7 Visibility requirements at a priority junction.

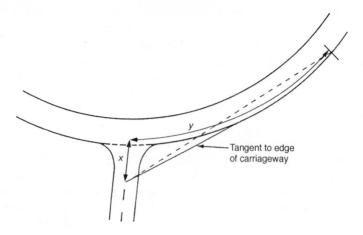

Fig. 7.8 Visibility requirements at a priority junction for a curved major road.

separate movements and reduce the number of simultaneous decisions required of drivers and pedestrians. For very heavy pedestrian movements formal pedestrian crossings are sometimes installed. The crossings should be located at a suitable distance from the main junction to ensure that drivers can see pedestrians clearly and queuing vehicles from the crossing do not seriously impede movement through the junction.

Pedestrian routes should avoid the most hazardous areas within the junction and should lead to suitably safe crossing points, with short crossing points that are not obstructed by parked vehicles. Pedestrian safety barriers (guard railing) and footway bollards can be used to prevent dangerous crossings and to guide pedestrians to the safety crossing points.

7.2.2   PRIORITY JUNCTION CAPACITY ASSESSMENT

The TRL computer program PICADY[5] (priority intersection capacity and delay) uses empirical formulae to model capacities and time-dependent queuing theory to estimate queues and delays for three- and four-arm major/minor junctions and left-right and right-left staggered crossroads. The program requires the input of demand traffic and pedestrian flows (where pedestrian crossings are present) and various geometric parameters which describe the junction. The program can simulate peak traffic conditions if only hourly traffic counts are available. The program can model the effects of flared minor roads and major road right-turning traffic blocking through traffic.

7.3   Roundabouts

The first roundabouts were constructed as circuses, with all vehicles entering the junction and turning left to merge with those circulating. There were no priority markings and vehicles were expected to 'zip' together by synchronising their speeds. Vehicles then circulated around the central island and diverged at their chosen exit. Two weaving manoeuvres were performed: one on entering the junction and the other just prior to the chosen exit. Under heavy flow conditions, roundabouts tended to lock, which occurred when drivers were prevented from leaving the junction by those entering. In the mid 1960s offside priority markings were introduced which reduced entry speeds, accidents and the tendency to lock.

The offside priority rule radically altered the method of operation of roundabouts. Weaving formulae for roundabouts[6] had produced very large junctions with wide straight links which encouraged high circulating speeds. Researchers at the Road Research Laboratory experimented with very small central islands.[7] Dramatic increases in capacity were achieved at smaller and smaller sites until the central island was reduced to a white painted circle. The entries were widened to allow for multilane approaches. The small size of the islands opened up many possibilities, and many small priority junctions, including staggered junctions, were converted with significant reductions in accident rates. At several sites, old-style roundabouts with long wide weaving sections were converted to multiple miniroundabouts, with traffic travelling in both directions around the central island (e.g. at Swindon and Hemel Hempstead; Figure 7.9).

However, the smaller islands led to problems with speeding vehicles on nearly straight paths through the junction. Recommended designs now incorporate speed reduction measures that include:

- deflection islands
- larger or domed central islands
- nearside kerb buildouts.

The designs of large roundabouts were also affected by the Road Research Laboratory (now TRL) research and most roundabouts are designed with circular central islands and flared approaches. Further research led to the capacity formulae used by the ARCADY computer program.[8]

The most important part of any roundabout design is ensuring that drivers approach it at suitable speeds. The DoT Advice Note TA 42/84[9] describes the entry curvature required to limit speeds, but earlier advice on preventing a see-through alignment is worth considering.

Speed reduction at small and miniroundabouts can be difficult to achieve as space is limited and vehicle swept paths have to be accommodated. Many small roundabouts have been equipped with abrupt kerb deflections that only serve to complicate the appearance and often contribute very little to speed reduction and safety.

The vertical alignment is often under the control of highway engineers who are highly skilled at providing the correct drainage cross-falls on roads and junctions. The old-style roundabouts were usually drained partly towards the central island and partly to the edge. If this method is adopted for small and miniroundabouts the roundabout can be hidden from the driver's view. On

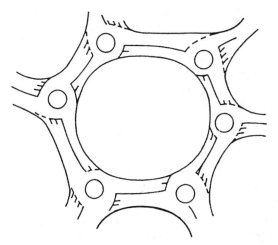

Fig. 7.9 Conversion of a conventional roundabout to a ring junction.

larger roundabouts the superelevation around the central island can encourage excessive speeds. Central islands should be raised in the centre of the junction and the carriageway drained towards the edges. A good reference for the design of miniroundabouts is *Mini-roundabouts – Getting Them Right!* by Clive Sawers.[10]

7.3.1   ROUNDABOUT CAPACITY ASSESSMENT

The TRL computer program ARCADY (assessment of roundabout capacity and delay) uses empirical formulae to model entry capacities and time-dependent queuing theory to estimate queues and delays for all types of single-island roundabouts. The program requires the input of demand traffic and pedestrian flows (where pedestrian crossings are present) and various geometric parameters to describe each roundabout entry and the overall size of the junction. The program can simulate peak traffic conditions if only hourly traffic counts are available. The program produces consistent results for conventional and small roundabouts, but when used for miniroundabouts it can underestimate potential capacity.

7.4   Signalled junctions

The general layout of signalled junctions is closely controlled by the proposed signal control method and the presence of pedestrian, cycle and bus facilities. It is normal to accept narrow, and sometimes substandard, lane widths at the junction stop lines. The advantage of multiple lanes often far outweighs the occasional problem with two or more lanes being dominated by a wide vehicle making a turn.

The requirements of pedestrians, cyclists and buses must be designed in from the beginning of the design process. They cannot be ignored and then tacked on to a near-complete design after detailed capacity assessments have been made.

In recent years the emphasis at traffic signals has moved away from pure handling capacity to provision of facilities for other users. In city centres pedestrian subways have lost favour and most new schemes include pedestrian crossings at ground level. Cycle advance areas are becoming more common and selective vehicle detection (SVD) for buses is in the process of becoming universal in many cities.

The increasing complexity requires careful attention to detail as there are more traffic islands being installed to accommodate ever increasing numbers of signal heads with pedestrian and cycle aspects. Large vehicle turning circles and swept paths must be rigorously checked. Very tight turns can slow large vehicles to the point where capacity is affected.

Carriageway and footway surfacing materials require careful consideration to complement the operation of the junction and to reduce the confusion to some road users. Tactile surfaces for pedestrians must be carefully placed to ensure that people with visual handicaps are helped safely through the junction. Traffic signal control is discussed more fully in Chapter 8.

References

1.   Department of Transport (1981 and 1984) *Road Layout and Geometry: Highway Link Design*, TD 9/81, and Amendment No1, HMSO, London.
2.   Department of Transport (1984) *Highway Link Design*, TA 43/84, HMSO, London.
3.   Department of Transport (1988) *TRACK prediction of vehicle swept paths*, Savoy Software.
4.   Department of Transport (1984) *Junctions and Accesses: the Layout of Major/Minor Junctions*, TA 20/84, HMSO, London.

5. Department of Transport (1985) *PICADY2: An enhanced program to model capacities queues and delays at major/minor priority junctions*, TRRL Research Report 36, TRRL, Crowthorne.

6. Wardrop, JG (1957) The Traffic Capacity of Weaving Sections of Roundabouts, Proceedings of the *First International Conference on Operational Research* (Oxford, 1957), pp 266–80, English Universities Press, London.

7. Ministry of Transport (1970) *Capacity of single-level intersections*, RRL Report LR356, TRRL, Crowthorne.

8. Department of Transport (1985) *ARCADY2: An enhanced program to model capacities queues and delays at roundabouts*, TRRL Research Report 35, TRRL, Crowthorne.

9. Department of Transport (1984) *The Geometric Design of Roundabouts*, TA 42/84, HMSO, London.

10. Sawers, C (1996) *Mini-roundabouts – Getting Them Right!*, Euro-Marketing Communications, Canterbury.

8

Signal Control

8.1 Introduction

Traffic signals are used to regulate and control conflicts between opposing vehicular traffic or pedestrian traffic movements. Without the use of signals at some sites the major flow would dominate the junction, making entries from the minor road impossible or very dangerous. At other sites the minor road might interfere with the flow of major road traffic to such an extent that excessive congestion would occur. Traffic signals cannot only improve junction capacity, but can also improve road safety.

Modern traffic signal controllers utilising microprocessors are reliable and flexible and can be programmed to handle multiple phases and other features, including:

- pedestrian facilities;
- predetermined fixed time;
- vehicle actuation including hurry calls actuated by excessive queues on certain arms, public transport or emergency vehicles;
- links by cables or cableless linking facilities to other nearby signalled junctions or pedestrian crossings and integration into an urban traffic control system (including adaptive systems such as SCOOT; page 82).

8.2 Fixed time control

In fixed time control the proportion of green time assigned to opposing arms is preset in accordance with historical traffic data. At most sites traffic flows vary throughout the day. Typically, inbound flows to a city are high in the morning peak period and outbound flows are high in the evening. Different green splits may therefore be required at different times of the day or different times of the year.

8.2.1 VEHICLE ACTUATION

Information about traffic demands on the approaches to signals is detected by inductive loop detectors buried in the road surface or by post-mounted microwave detectors. The signal controller can extend the green stage on the relevant arms from a preset minimum to a maximum. These systems are highly responsive and can minimise delays and maximise capacity at isolated independent sites. Hurry calls, to prevent excess queuing or to assist emergency vehicles and

Distance

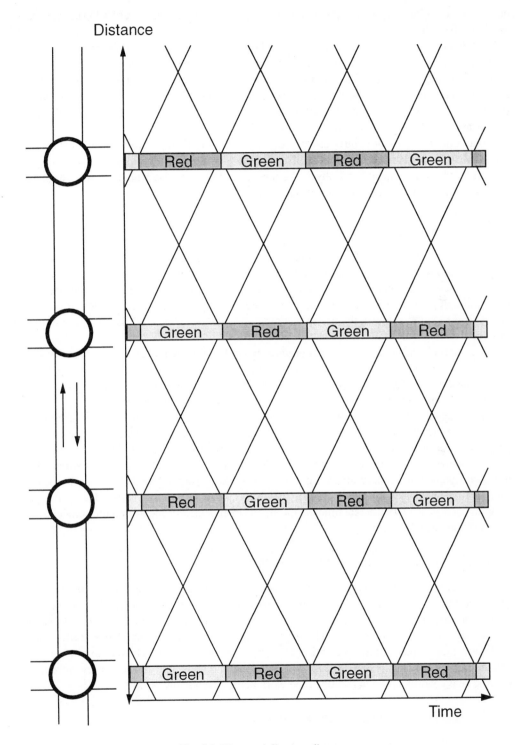

Time

Fig. 8.1 Time and distance diagram.

buses, are very effective. Time savings for buses fitted with transponders at SVD sites can be as much as 10 seconds per bus.

Neighbouring traffic signals can be connected by cable links to provide uninterrupted progression for vehicles on the main route. The difference between the start of the green stage and the downsteam junction is related to the journey time at cruise speed. The difference in the start of the green stages is known as the offset. Successive sets of signals can also be linked to produce a steady progression along a road. Signal offsets can be calculated or arrived at graphically using a time and distance diagram (Figure 8.1).

Modern microprocessor controllers are equipped with digital time clocks which can be very accurately set using hand-held terminals. This facility allows signal timings and offsets to be synchronised without the need for a physical connection. However, interrupted power supplies or failure of a component can cause synchronisation to be disrupted.

8.3   Urban traffic control

An extension of the linking and coordination of adjacent signals to a whole network of signals can reduce journey times and congestion by maximising the network's capacity. Networks of signals are much more complex than simple linked signals as there may be more than one major route through the network and these routes may not be easy to identify. As predetermination of effects is difficult to achieve, the use of a computer model of the network is needed to optimise the signal settings for all junctions, such as TRANSYT.[1]

Urban traffic control (UTC) systems produce many benefits by increasing the capacity of the whole network and reducing travel times, fuel consumption and, hence, air pollution. Unfortunately this additional capacity is quickly filled by additional traffic and, apart from the steadier progression through the network, similar levels of congestion return.

Clearly, any additional capacity created by urban traffic control systems must be used wisely and a strategy for its use must be agreed before implementation. Spare capacity can be utilised to provide bus priority measures, cycle facilities, pedestrian crossings and to alleviate rat-running through residential and shopping streets. The Transport Research Laboratory has estimated that a well-maintained fixed time urban traffic control system using TRANSYT plans can produce journey time benefits of around 15% and a SCOOT[2] fully adaptive system around 20%. Adaptive systems such as SCOOT use real time data from on street detectors and a computer model to continuously update signal timings.

8.3.1   FIXED TIME URBAN TRAFFIC CONTROL

Large urban traffic control systems are controlled by a central computer which initiates signal plans for varying flow patterns. The plans are routinely activated by a clock within the controlling computer but can be overridden manually; special plans can be programmed in to accommodate unusual or special events. The greatest problem with fixed time urban traffic control systems is that the plans need to be updated regularly to keep pace with changes in traffic pattern. This represents a considerable expense in organising and collecting traffic survey information and the heavy workload often results in the work being given a low priority.

8.3.2   ADAPTIVE, TRAFFIC-RESPONSIVE UTC SYSTEMS

Adaptive systems require three main components:

- vehicle detectors

- central controlling computer
- implementation of signal settings within the traffic signal controller.

Data are transmitted to and from the street over public telephone lines, purpose built communication systems or, occasionally, radio links.

8.3.3   PLAN SELECTION

Traffic flow data are used to select appropriate, predetermined fixed time signal plans. This is similar to fixed time urban traffic control systems that use the time of day for plan selection. The fixed time plan chosen will always be a compromise and each plan must be prepared using a tool such as TRANSYT. The SCATS system developed in Australia, for use in Sydney, is a successful example of this type of system.

8.3.4   FULLY ADAPTIVE SYSTEMS

Fully adaptive systems use traffic flow data to calculate optimum signal settings continuously and adjust signal settings in a controlled manner. Early adaptive systems suffered from a number of problems such as slow response, inadequate prediction and too-frequent plan changing.

The most widely used fully adaptive system in the United Kingdom is SCOOT (split cycle time and offset optimisation technique). SCOOT is the result of cooperation between the three main traffic signal supply companies in the UK, the Transport and Road Research Laboratory (TRRL, now TRL) and the Department of Transport.

SCOOT uses vehicle detectors (usually inductive loop detectors cut into the road surface) placed at some distance from the signal stop line. Correct siting of detectors is critical to the success of a SCOOT system. The detectors should be placed as far as possible from the stop line. There should be minimal change in traffic flow between the loop and the stop line and should be at least 10–15 m downstream of the preceding upstream junction. Figures 8.2–8.4 show typical SCOOT loop sites and dimensions.

There should be at least one loop installed for each signal stage. Additional loops can be used for right turn stages and for major flow sources. For example a car park exit between the main loop and stop line could be provided with a loop (Figure 8.5).

Loop occupancy is checked four times per second and is used to provide information on traffic demand, queues, congestion and exit blocking. The unit of flow and occupancy used for SCOOT loop detectors is the link profile unit (LPU). For comparison purposes there are 17–18 LPUs per vehicle. (A saturation flow of 2000 vehicles per hour is equivalent to a SCOOT saturation occupancy of 10 LPUs per second, a 10 vehicle queue equates to 170–180 LPUs.)

Data from the detectors is transmitted via an outstation transmission unit (OTU) to an instation transmission unit (ITU) in the SCOOT control room. The SCOOT computer runs an on-line traffic model of the network, or networks, under its control – a database containing information about the junctions being controlled, the loop detectors and their stop line distances and relationship to the junction. The traffic model calculates the optimum signal settings and then transmits the new timings to the individual signal controllers on-street.

The optimisers within the SCOOT model produce a large number of very small changes. Large changes can be very disruptive to traffic flow and can confuse drivers.

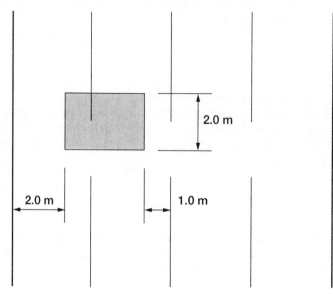

2.0 m

2.0 m   1.0 m

Two lanes in each direction

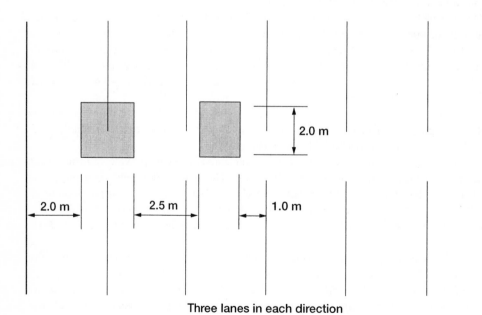

2.0 m

2.0 m   2.5 m   1.0 m

Three lanes in each direction

Fig. 8.2 Loop detector positioning for SCOOT.

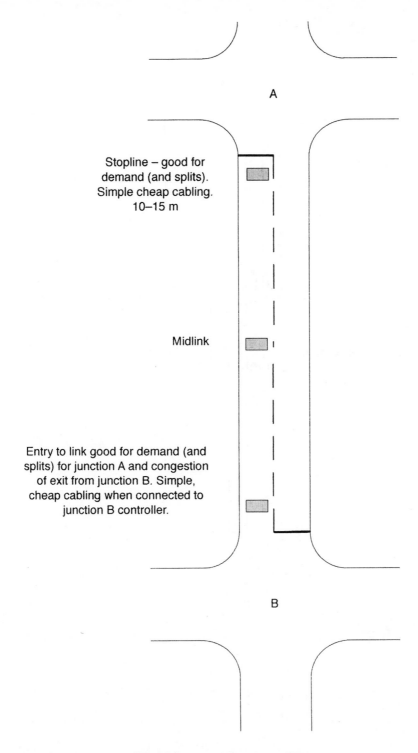

A

Stopline – good for
demand (and splits).
Simple cheap cabling.
10–15 m

Midlink

Entry to link good for demand (and
splits) for junction A and congestion
of exit from junction B. Simple,
cheap cabling when connected to
junction B controller.

B

Fig. 8.3 Loop positions along a link.

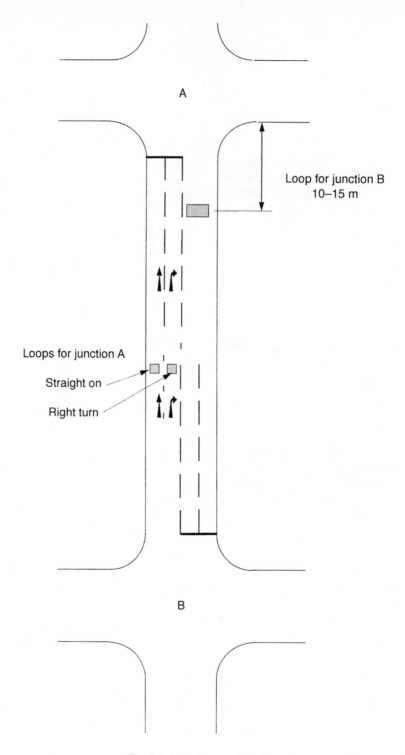

A

Loop for junction B
10–15 m

Loops for junction A

Straight on

Right turn

B

Fig. 8.4 Loop detector locations.

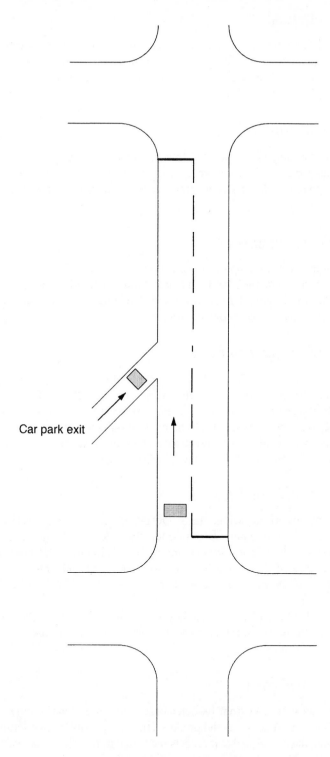

Fig. 8.5 Link with car park exit.

8.3.5   SPLIT OPTIMISER

The aim of the split optimiser is to minimise delay by minimising the degree of saturation and congestion. Each junction is optimised in turn, 5 seconds before every stage change. A maximum of 4 seconds temporary advance or retard is permitted with a permanent 1 second change within the predetermined maximum and minimum green times.

8.3.6   OFFSET OPTIMISER

The offset optimiser aims to smooth the flow through successive junctions. All upstream and downstream links for every junction are optimised for every cycle. Where links are very short they can be equipped with fixed offsets; offsets can also be weighted or biased for specific purposes such as bus priority.

8.3.7   CYCLE TIME OPTIMISER

The objective of the cycle time optimiser (CTO) is to achieve 90% target saturation. Cycle time is adjusted every 2.5 or 5 minutes between a preset minimum practical and maximum range in 4 second steps. The whole network runs on the same cycle time or half cycle time for certain double cycle time junctions and pedestrian crossings.

8.3.8   SELECTIVE VEHICLE DETECTION

Selective vehicle detection (SVD) has the potential to reduce bus delays at independent junctions by 32% (22% in UTC networks) and reduce variability by 22%. SVD uses detectors, usually loops cut into the road surface and bus-mounted transponders, to advance or extend traffic signal stages to provide priority for buses. The efficiency of SVD is dependent upon the precise location of the detectors, in relation to the junction, and almost instantaneous transmission of a vehicle's presence to the signal controller or the SCOOT computer.

8.3.9   SVD DETECTOR LOCATION

To obtain effective bus priority within the SCOOT-UTC areas, all approaches to signalled junctions with bus services will require detectors. Ideally, detectors should be sited downstream of the nearest bus stop and, normally, 70 m from the signal stop line. However, useful benefits for buses can be obtained if this distance is no less than 35 m and limited benefits can still be obtained below 35 m. On higher-speed roads and in free flow conditions the optimal distance may increase up to 120 m.

Loop detectors should be placed so that they are not activated before a bus priority sequence is needed. Therefore, the precise location should take account of the normal stopping place of buses (Figure 8.6).[3]

8.3.10   SCOOT VALIDATION

The SCOOT model uses data from its detectors to optimise signal settings and the resulting queue lengths. To ensure that the model produces the best possible signal settings, it is necessary to tune or calibrate the model so that it reflects the actual traffic conditions as closely as possible. Once validated there should be no need to revalidate, unless there is a major change in site conditions or traffic composition. The following changes at a site would create a need to revalidate:

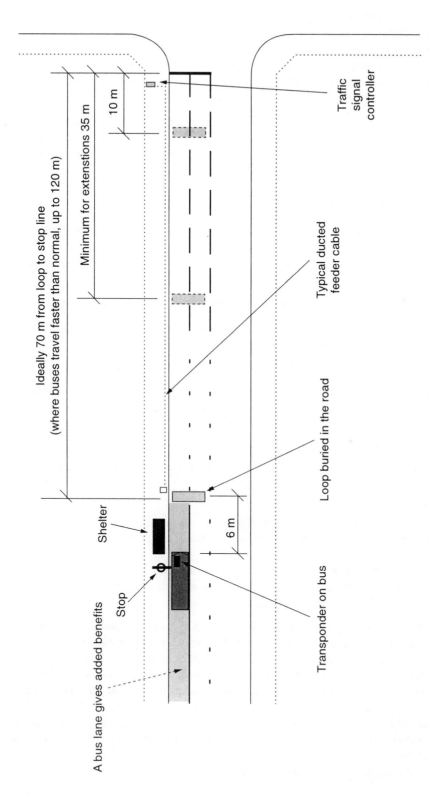

Fig. 8.6 Bus loop detector siting.

Ideally 70 m from loop to stop line
(where buses travel faster than normal, up to 120 m)

Minimum for extenstions 35 m

10 m

Traffic
signal
controller

Typical ducted
feeder cable

Loop buried in the road

Shelter

6 m

Stop

A bus lane gives added benefits

Transponder on bus

- loop resiting;
- change in junction geometry;
- addition or deletion of traffic flow source;
- change in speed or parking regulation;
- change in composition of traffic, vehicle size or performance;
- technological changes to vehicles or control systems.

Validation requires a team of two people on site, usually a traffic engineer and a technician, and two-way voice communication with the control centre. It is also possible to use a hand-held terminal attached to the OTU in the signal controller to monitor SCOOT input/output messages. Approximately one day is required to validate a typical junction.

8.3.11 SCOOT VALIDATION PARAMETERS

- Journey time – the time taken by an average vehicle in a free flowing platoon between the detector and the stop line.
- Saturation occupancy (SatOcc) – maximum outflow rate of a queue over the stop line in LPUs per second:

  SatOcc (LPUs per second) × 200 = saturation flow in vehicles per hour

- Maximum queue – average largest number of vehicles that can normally be accommodated between the detectors and the stop line.
- Green start lag – intergreen plus 1 or 2 seconds.
- Green end lag – 2 or 3 seconds.
- Main downstream link – determined from local knowledge and observation.
- Default offset – journey time less standing queue clearance time.

8.3.12 SCOOT DATA

The SCOOT model requires input data to be held in a structured database, with the following hierarchy:

- area;
- region – part of an area for which there is a common cycle time;
- node – individual signal controlled junctions or pedestrian crossings;
- stage – junction signal stage;
- link – normal, entry, exit, filter or bus priority link;
- detector – detector identity (there may be more than one detector on a link and these should be given separate identities).

SCOOT can be used to favour certain links or series of links but overall network delay will increase. Occasionally users of a SCOOT network observe that a particular junction is not operating at capacity. It should be remembered that SCOOT is optimising a whole network and that the individual junction is an integral part of that network. It is possible that, if the junction is altered in isolation, queues will build up on another part of the network resulting in an increase in overall network delay.

An associated program is ASTRID, which can be used to extract historical data from the SCOOT database. This information is very accurate and can be obtained for extended periods. Data is supplied in LPUs and then converted to vehicles using the standard 17–18 LPUs per vehicle or calibrated by conventional traffic counts.

8.4   Other urban traffic control facilities

8.4.1   FAULT MONITORING

Certain fault conditions can be transmitted directly to the control centre and used to direct maintenance operations.

8.4.2   CAR PARK INFORMATION AND VARIABLE MESSAGE SIGNS

Additional detection can be used to monitor car park arrivals and departures. Space availability is then displayed on variable message signs (VMSs) installed at key locations.

8.4.3   HURRY CALLS AND GREEN WAVES

Excess queues can be dissipated when queue detectors are occupied at important sites. Priority routeing for emergency vehicles can be implemented using predetermined plans or by using transponders fitted to the vehicles and special detectors. Transponders are commonly fitted to buses for conventional SVD junctions and more recently in fixed time urban traffic control and SCOOT areas.

8.5   Traffic signal capacity assessments

8.5.1   INDEPENDENT AND ISOLATED JUNCTIONS

In recent years there has been a rapid increase in the number of traffic signalled junctions, either linked to adjacent junctions or under urban traffic control. However, the majority of signalled junctions operate independently of others. This has the advantage that cycle times and green splits can be optimised to a high level without the need for compromises demanded by nearby junctions. The traffic engineer must be able to assess the likely queues and delays when changes are being made to existing junctions or new installations are being considered. The assessments can be divided into two parts:

- capacity, queues and delays
- signal settings or timings.

The capacity of a junction is dependent on the green time available on each arm and the maximum flows over the signal stop lines (saturation flow). For a single traffic stream during one cycle of the signals the capacity is:

$$\mu = \lambda S$$

where

μ = maximum flow rate over the stop line/unit time;
λ = proportion of the cycle which is effectively green;
S = saturation flow for the approach.

Traffic flow is expressed in passenger car units per hour (PCU/h). PCU:vehicle ratios for different vehicle types at traffic signals are:[4]

cars and light goods vehicles	1.0
medium goods vehicles	1.5
heavy goods vehicles	2.3
buses and coaches	2.0
motorcycles	0.4
pedal cycles	0.2

Saturation flow occurs when there is continuous queuing on the approach and is determined, at existing junctions, by direct measurement or from empirical formulae derived from public road and track tests carried out by the TRL.[5-7]

These formulae predict saturation flows from measured geometric parameters at the junction as follows:

- lane width
- gradient
- lane position (i.e. nearside or non-nearside)
- vehicle type composition
- turning radius (as appropriate).

Effective green time is used to simplify the typical flow profile (Figure 8.7) which assumes that there is a queue on the approach before and after the end of the green stage.

The practical capacity of an approach or stream is:

$$p \frac{sg}{c}$$

where

g = effective green time
c = cycle time
p = maximum practical degree of saturation (usually taken as 0.9)
s = saturation flow.

All approaches should operate within their practical capacities or the junction will be overloaded. If the demand flow is less than the practical capacity the differences are termed the reserve capacity.

There are a number of computer programs available for the assessment of independent signalled junctions. The most popular in current use are OSCADY[4] (optimisation of signal capacity and delay) and LINSIG.[8]

OSCADY was developed by the TRL and optimises signal timings for a period of the day with varying demand flows and calculates the resulting queues and delays. Queues and delays are calculated using time-dependent queuing theory similar to that used in the ARCADY and PICADY programs.[9] OSCADY also produces signal settings but LINSIG has gained popularity with traffic signal engineers in recent years.

TRANSYT (traffic network study tool) was developed by TRL to optimise timings for a network of traffic signals. The resulting timings are used for fixed time urban traffic control systems and to provide initial settings for SCOOT.

TRANSYT models the dispersal of platoons of vehicles leaving upstream junctions as they travel towards the next junction and automatically selects the best timings for the network. Priority junctions and roundabouts can be modelled reasonably well if they occur within the network, but the program is intended primarily for fully signalled networks. The network is represented by a series of nodes (junctions) and links (traffic streams). A single link can represent several traffic streams at a stop line or several links can share the same stop line. Stop line saturation flows, link lengths, link cruise speeds and demand flows are supplied to the model.

The model calculates a performance index (PI) based on the monetary value of delays and stops for each set of trial signal settings. The optimiser searches for the minimum PI for the network.

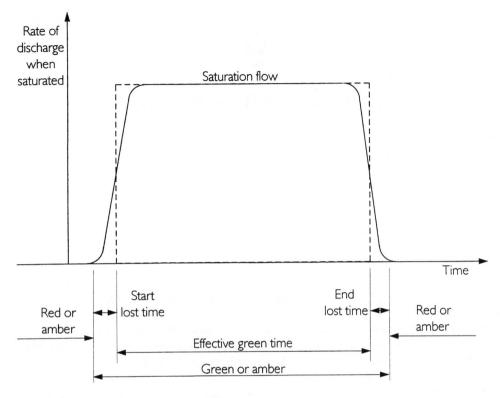

Fig. 8.7 Saturation flow profile.

References

1. Department of the Environment (1980) *User guide to TRANSYT version 8*, TRRL Laboratory Report 888, TRRL Crowthorne.
2. Department of Transport (1981) *SCOOT: a traffic responsive method of coordinating signals*, TRRL Laboratory Report 1014, TRRL, Crowthorne.
3. Traffic Control Systems Unit (1997) *Bus priority – selective vehicle detection in London*, (unpublished, may be available on request), London Transport Buses, London.
4. Department of Transport (1987) *OSCADY: A computer program to model capacities queues and delays at isolated traffic signal junctions*, TRRL Research Report 105, TRRL, Crowthorne.
5. Department of Transport (1986) *The prediction of saturation flows for road junctions controlled by traffic signals*, TRRL Research Report 67, TRRL, Crowthorne.
6. Department of Transport (1982) *Traffic signalled junctions: a track appraisal of conventional and novel designs*, TRRL Laboratory Report 1063, TRRL, Crowthorne.
7. Webster, FV and Cobbe, FM (1966) *Traffic signals*, Road Technical paper 56, HMSO, London.
8. Simmonite, B (1985) LINSIG: A program to assist traffic signal design and assessment, *Traffic Engineering and Control*, **26**(6).
9. Kimber, RM and Hollis, EM (1979) *Traffic queues and delays at road junctions*, TRRL Report LR 909, TRRL, Crowthorne.

9
Parking: Design and Control

9.1 Introduction

The traffic engineer will need to know how best to provide parking, and how to control parking facilities, both on the highway and off street, both in surface sites and structures. This chapter describes the key factors that need to be considered.

Parking provision on the highway in England, Scotland and Wales is constrained by legislation. Government rules and guidelines determine where parking can be provided, the methods of control and the design standards to be used. Separate legislation applies in Northern Ireland.

Off-street car parks are provided to meet a variety of needs and the type of need can affect the design of the car park. For example, an office car park could be designed to a lower standard than would be considered necessary for a public car park, as most of the users will be familiar with the geometry of the car park. If, for example, cars are parked too close together, a work colleague can usually be found and asked to move their vehicle.

Although off-street car park design is not governed by legislation in the same way as parking on the highway, the operation of a public car park can be subject to regulation using statutory powers.

Car parking control equipment is becoming increasingly more sophisticated, and the advent of microprocessor-based systems has allowed parking control systems to become more flexible, to meet the varied demands of users more closely. However, as the availability of increasingly sophisticated control systems has affected the way parking is controlled off-street, on-street parking controls have seen a different kind of revolution. This has allowed more flexibility, to better meet users' needs, through the use of both high-tech and very low-tech control equipment.

9.2 On-street parking

Most roads, in most places, are not subject to any form of parking control. It is widely believed that there is a right to park where no controls are present. This is not true. There is a right to pass along (travel) on a highway but no absolute right to stop. Indeed, as far back as 1635, parking problems in London had become so bad that a Royal proclamation was issued which declared that:

> We expressly command and forbid, that no hackney or hired coach be used or suffered in London, Westminster or the suburbs thereof, ...

Earlier still in ancient Rome, things reached such a state that in an attempt to restore order the death penalty was threatened to anyone leaving a chariot parking on a public thoroughfare.

Generally speaking, a vehicle parked at the kerbside, providing it is not parked dangerously or blocking traffic, will not attract police action although, in theory, the driver could be prosecuted for obstruction.

Formal parking on the highway can be provided for general use, or for particular groups of users. For example, kerbside parking could be set aside for particular classes of vehicles; this could include:

- all vehicles below a certain size (i.e. excluding heavy commercial vehicles)
- solo motor cycles
- taxis
- buses
- vehicles belonging to registered disabled
- local residents' vehicles
- local businesses
- doctors
- diplomats
- police vehicles
- specialist vehicles, such as a mobile library.

The provision can be made all the time (at any time), on certain days or at certain times of day and the provision can be free or be charged for. Thus, generally speaking, a taxi stand is available only to taxis at any time. However, other facilities may only operate part time. For example, parking for residents in a city centre may only be protected during the working day, typically 08.30–18.30 hrs, Monday–Friday, although the increasing tendency for late night opening and the introduction of Sunday trading means that the timing of many of these regulations will increasingly be rethought.

Where activity is intermittent, for example near a football stadium, restrictions may not be appropriate all the time and restrictions may only apply on match days. This type of restriction is increasingly common in areas where a local community requires occasional protection from a short-lived sudden influx of parkers.

9.2.1 ROAD MARKINGS

The design and marking of parking bays on the highway is governed by legislation, with approved markings shown in Schedule 6 of the *Traffic Signs Regulations and General Directions (TSRGD) 1994.*[1] These show a variety of markings for both cars and for specialist bays, for buses, taxis etc; see Figure 9.1. Echelon parking, as shown in Fig. 9.1(e), is seldom used because, although it may allow additional parking spaces to be accommodated in a wide street, drivers tend to have difficulty manoeuvring in and, particularly, out of the spaces.

The current regulations specify a standard bay width, of between 1.8 and 2.7 m, although a bay width of 2.7–3.6 m is specified for parking bays for the disabled. In reality, these widths may not be achievable in narrow streets, and it may be necessary to make a pragmatic trade-off between the need to provide parking and the need to ensure sufficient road width for moving traffic. Prior to the 1994 regulations, government guidance allowed bays as narrow as 1.6 m. Below this, the width of the average modern car would mean that many vehicles would protrude outside the bay.

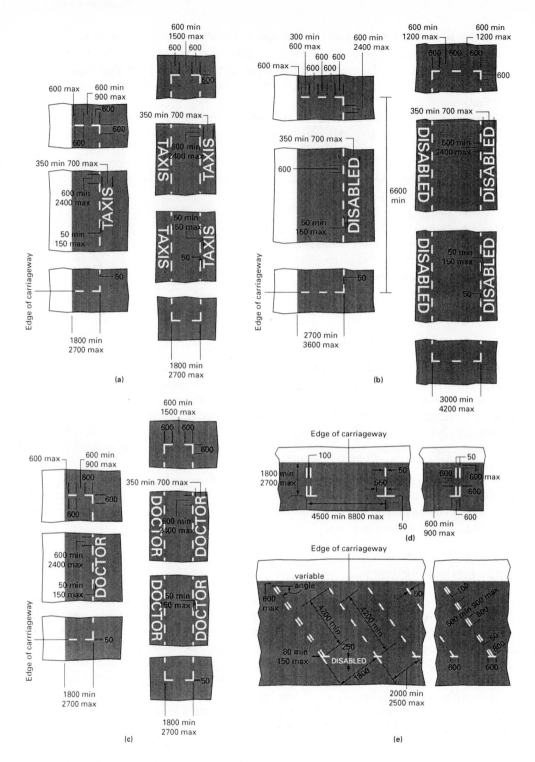

Fig. 9.1 Road markings for special types of parking.[1]

9.2.2   *SIGNING*

Where parking is provided with restrictions then signs are used to show what type of vehicle can park and how. These regulatory signs are specified in Schedule 2 of the *TSRGD 1994*. The signs either specify the type of vehicle allowed to use the parking place, or specify the conditions and restrictions attached to the use of the parking place. If the parking is available for a particular class of vehicle to use, without other constraint, the signing will identify the class of vehicle that can use the parking, as shown, for example in Figure 9.2. Alternatively, the signing may display information about the limits and conditions of use for parking. Examples of this type of parking signing are the diagrams shown in Figure 9.3.

9.2.3   *REGULATIONS*

Parking on the highway can be provided using a number of different methods of regulation. The principal legislation is the Road Traffic Regulation Act 1984,[2] as modified by later legislation.

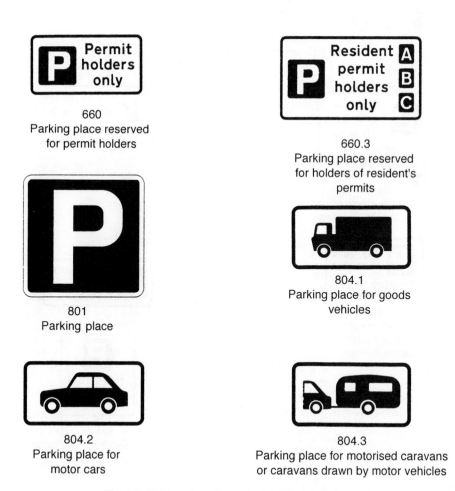

660
Parking place reserved
for permit holders

660.3
Parking place reserved
for holders of resident's
permits

801
Parking place

804.1
Parking place for goods
vehicles

804.2
Parking place for
motor cars

804.3
Parking place for motorised caravans
or caravans drawn by motor vehicles

Fig. 9.2 Parking signs for particular types of road users.[1]

Part I of the Act provides general powers to allow traffic regulation. Using powers in this part of the Act, a highway authority can make regulations to restrict parking or, within the wording of the legislation, 'waiting and loading'. This legislation also allows for exemptions to be made to any order and so some parking provision made be made by exception. This means that, in a street where parking is otherwise banned, a parking place may be provided in this way, for example for

660.5
Parking place reserved for
voucher parking during the
period indicated

661.1
Restrictions on length of waiting
time and return period

661.2
'Pay and Display'
ticket-regulated place

661.3
Location of
'Pay and Display'
ticket machine

661.4
Drivers must obtain and display
parking tickets
(Alternative types)

Fig. 9.3 Parking restriction signs.[1]

motorcycles. Thus if any other vehicle parks in a motorcycle bay it is in breach of the more general order banning waiting. Great care has to be taken in using the law in this way. If a street had a waiting restriction, without a ban on loading, with, say, a motorcycle parking bay, then any other vehicle parking in the bay would be committing an offence but, if the vehicle were loading goods, then no offence would be committed! This is clearly confusing and unsatisfactory.

Part IV of the Act provides powers to create parking places and can be used to create specific parking places on the highway. Using these powers, the highway authority can make explicit provision for parking in a street, either free or for payment. Part I powers would not be used, for example, in a street to provide a disabled bay where no other controls were proposed. In these circumstances, the bay would be specified, using the powers contained in Part IV of the Act.

In some towns and cities, where parking controls are needed on an area-wide basis, the required provision is made by creating a controlled parking zone (CPZ). Within a CPZ all the streets can be controlled by an overall order which is applied to all the streets within the area. Normally this is a restriction on parking for part or all of the day. The powers in Part IV of the Act are then used to specify places where parking is allowed by exception. Typically these powers would be used to specify parking for residents, plus parking for visitors to the area. The logic of a CPZ order is to have an overall order which applies to all the streets, with specified areas then set aside for parking. This has the advantage that, if a parking place is no longer needed, then the underlying restriction automatically applies as soon as the parking is removed.

The statutory responsibility and power for introducing street parking controls resides with the highways authority. The procedures that a highway authority has to follow in seeking to introduce controls are set out in the traffic order procedure regulations and these are contained in two documents:

- *Local Authorities' Traffic Order (Procedure) (England and Wales) Regulations 1989*, SI 1989/1120;[3]
- *Local Authorities' Traffic Order (Procedure) (England and Wales) (Amendment) Regulations 1993*, SI 1993/1500.[4]

9.3 Off-street car parking

The design of a car park structure is a complex exercise requiring a detailed understanding of many factors. This book cannot hope to provide more than the basic issues that anyone embarking on this exercise needs to address.

The layout of a surface car park will be greatly influenced by the shape and form of the land over which it is constructed. Obviously a flat, square site offers an easier design problem than a steeply sloping irregularly shaped one. As a rule of thumb, to estimate the number of cars that can be accommodated on a site, each car space requires about 20–25 m^2 of space.

The basic element in a car park is a rectangular parking space which would typically be of the order of 4.75 m by 2.4 m. However, bays could be as small as 4.6 m by 2.2 m, where space is at a premium, and widths could be up to 3.6 m, or more, for disabled parking. Generally speaking, short-stay car parking, where there is a constant coming and going, operates more efficiently with wide bays, and the Institution of Highways and Transportation (IHT) recommend a 2.5 m bay width for short-stay car parking.[5]

Parking bays are grouped together in rows and parking places at 90° tend to provide the most efficient use of a site, although, if the car park shape is irregular, echelon parking may be desirable to use the space more efficiently. Echelon parking may also be desirable in situations of very short-stay parking where dynamic capacity, being the maximum flow of traffic along aisles, is more important than static capacity.

The rows of parking have to be segregated by aisles to allow vehicles to access the bays, and the aisle width depends on whether it is carrying one- or two-way traffic. Opinions are mixed on the minimum required. A two-way aisle needs to be 6.95 m according to the IHT, with a 6 m aisle being acceptable for one-way traffic. Other sources recommend that a 6 m aisle is sufficient to carry two-way traffic.

Designing the physical layout of a car park should go beyond the engineering design of the parking spaces to include consideration of the aesthetic and environmental design, and the creation of a secure environment for cars and drivers. A system of car park design assessment has been created under the *Secured by Design* scheme promoted jointly by the Association of Chief Police Officers and the Automobile Association.[6] The scheme considers factors such as:

- levels of lighting;
- landscaping and the extent to which landscaping offers places where criminals could hide – the guidelines recommend that planting should be below 1 m or above 2 m;
- levels of natural surveillance, that is whether the car park can be seen by people passing by;
- the level of security offered by CCTV or by patrolling staff.

There are a variety of designs for car parking structures. The floor of the car park can be flat with connecting ramps. Ramps connecting floors can be straight or can be helical. The floor can be built on a gradient to obviate the need for connecting ramps, in effect creating a continuous spiral.

Where land is at a premium, mechanical parking systems can be used. These car parks typically use palletised systems to stack cars and require less space per car as there is no need to allow manoeuvring space.

9.4   Disabled drivers

Parking facilities designed specifically for, or to include, disabled drivers should be fully accessible throughout for wheelchair users. Parking bays should be 1.5 times the width of a normal bay with access by ramps with a maximum gradient of 7%. Where lifts are provided they should have controls at wheelchair level, that is at about 1.2–1.4 m.

9.5   Servicing bays and lorry parking

Shops and factories require service bays, to accommodate commercial vehicles delivering and collecting goods. The demand for service bays should be determined as part of a traffic/transport impact study, undertaken to assess the transport needs of a new development.

A loading bay would typically be 3.3 m wide with a length determined by the scale of vehicle expected to use the bay, which could be up to 15.5 m long. The access route to the loading bay should be sufficient to allow the commercial vehicle to access the bay and manoeuvre into and out of the bay with ease. A 15 m articulated lorry typically has a 13 m radius outer and a 5.3 m inner turning circle and the design of the manoeuvring area for service vehicles should ensure that this swept path is clear of obstructions.

This must be approached with common sense. Lorries are driven by human beings and allowance must be made for the variability that this will introduce to the path that a lorry follows when manoeuvring. Thus it is not good practice, having determined the path of the vehicle, to place an obstruction within a few centimetres of the theoretical limits of the swept path. It is surprising how often otherwise sensible architects and engineers do just this.

9.6 Parking control systems

Parking on the highway can be controlled in a number of ways by regulation. The regulation can define:

- the days, or times of day, that vehicles can park;
- the length of time that a vehicle can park;
- the type of vehicle that can park;
- the charge for parking.

The responsibility for this regulation rests with the highway authority which has to make a Traffic Regulation Order setting out the form of control required. The primary powers which enable the highway authority to control parking are set out in the *Road Traffic Regulation Act 1984*,[2] which has subsequently been amended by various enactments and statutory instruments. The main legislation is summarised in Table 9.1.

Table 9.1 Summary of main legislation in the *Road Traffic Regulation Act 1984*

Enactment	Section	Application
Road Traffic Regulation Act 1984	1	Prohibiting waiting and loading outside London
	6	Prohibiting waiting and loading in London
	9	Making experimental orders
	32–42	To provide free-parking on-street and off-street parking places
	43–44	Local authority powers to license public off-street parking places
	45–56	Parking on the highway for payment
	61	Loading areas
	63	Cycle parking
	95–121	Offences
	122	Exercise of powers
Local Government Act 1985		This legislation abolished the GLC and the other Metropolitan councils and amended the 1984 Act to pass the relevant powers to the successor bodies
Road Traffic Regulation (Parking) Act 1986		This act amended the 1984 Act to allow the use of cashless payment systems
Road Traffic Act 1991	41	Simplification of procedures for changing off-street parking charges
	42	Simplification of procedures for changing on-street parking charges
	43	The power to create permitted and special parking areas outside London
	44	Parking attendants
	63–82	Decriminalised parking powers for London

9.6.1   FREE PARKING

As car use and hence parking demand grows, the need to control parking increases. The simplest form of control is a system which seeks to limit how long a vehicle can park. In Great Britain this can be done using a regulation made under Section 35 of *The Road Traffic Regulation Act 1984* (RTRA 1984) to set a limit on the length of stay. This regulation is indicated using the sign shown in Figure 9.4 which requires that the vehicle, once parked, cannot return within a certain period of time. Enforcement of this system relies on parking enforcement staff recording the details of each vehicle they see parked and returning sufficiently frequently to identify vehicles that have stayed beyond the maximum stay, but have not had time to leave and legally return.

Thus, if the limit was 1 hour, with no return for an hour, the enforcement staff need to visit more frequently than once every 2 hours and less frequently than once an hour. This would ensure that, if they saw a vehicle parked in the same place on successive visits, it must have committed an offence as it could not have left and returned after being away for the required hour.

Clearly, to ensure that this type of parking control works it is necessary to have either a high level of compliance, or a high level of enforcement. For the enforcement to be effective the patrol staff have to keep good records, making the system labour intensive and hence costly to run.

In some places where free parking is still available parking is controlled using a parking disc (Figure 9.5). The disc, which is described in BS 6571 Part 8,[7] is a simple cardboard clock which

661.1
Restrictions on length of waiting
time and return period

Fig. 9.4 Free parking restriction sign.

662
Period during which waiting is limited and display
of a disc is required at a parking place in a disc zone

Fig. 9.5 Disc zone parking restriction sign.

requires the driver to set the time of arrival. This simplifies the enforcement process as now all the enforcement officer has to do is read the clock face to see if the vehicle has overstayed. The driver commits an offence by over-staying or by not having a disc. Discs are also used by disabled drivers where their stay is time limited.

There are many towns where parking is still free, using one of these systems. However, as car ownership and use increases, local councils are tending to move towards systems where a driver has to pay for parking.

9.6.2 CHARGED-FOR PARKING

The reasons for introducing parking charges can be complex and include:

- the need to deter parkers;
- the need to cover the costs associated with providing the parking;
- wider transportation policy objectives.

Whatever the reason, the decision to introduce charges will mean that the highway authority will have to make a traffic order, to bring into effect the charges and introduce a new form of

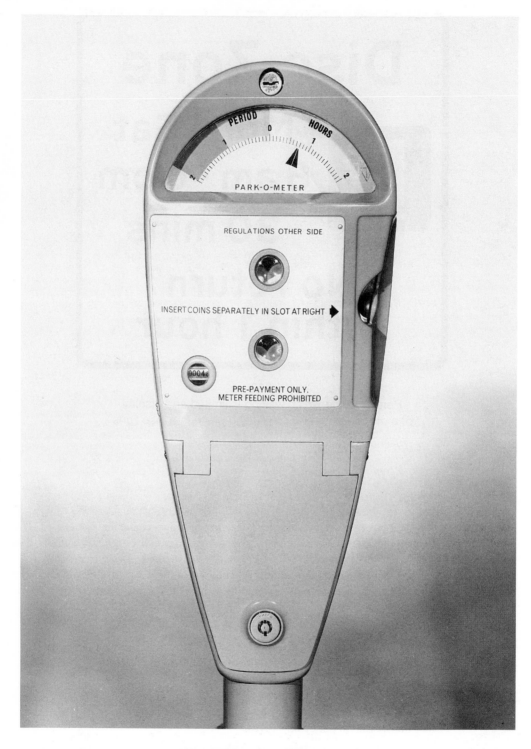

Fig. 9.6 Clockwork parking meter.

control in order to collect the charges. As part of the decision to introduce charges, the authority may also decide to allocate parking space to certain groups, with spaces set aside for particular groups of users. Thus a council could set aside some spaces for local residents, some for local businesses, some for motorcycles and some spaces for the use of disabled drivers, with other spaces for all-comers.

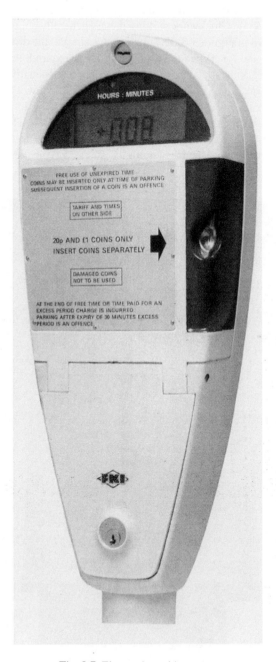

Fig. 9.7 Electronic parking meter.

Charges for on-street parking can be controlled a number of ways. The most common ones are:

- clockwork parking meters
- electronic parking meters
- pay and display
- parking vouchers.

A parking meter is used to control an individual parking bay, and allow a driver to pay for their parking at the time of parking. Clockwork parking meters generally conform to BS 6571 Part 1;[8] a typical example is shown in Figure 9.6. The operation is simple: the driver inserts coins and the meter registers the time paid for. When the time paid for has expired, the meter shows a flag in the display to show to a parking attendant or traffic warden that a penalty has been incurred.

In electronic meters, described in BS 6571 Part 2,[9] the clockwork mechanism has been replaced by an electronic timer, display and coin validation system. In addition to offering the greater accuracy and reliability inherent in a electronic mechanism, the electronic coin validation system allows better checking of coins and for a greater variety of coins to be used than would be practical with a purely mechanical device. Figure 9.7 shows a typical example of an electronic parking meter.

Fig. 9.8 Pay and display parking meter.

A pay and display machine allows a number of spaces to be controlled using a single machine. The number of bays controlled will depend on the layout of the streets and the closeness of the bays to each other. Like a parking meter, the driver has to pay for parking on arrival. However, since the machine controls a number of bays, the parker will have to walk to the machine to pay. The pay and display machine issues a ticket which the driver displays within the car, so that it can be checked by the patrol staff. Figure 9.8 shows a pay and display parking meter conforming to BS 6571 Part 3.[10]

Pay and display machines are electronic and can be powered by either batteries or from the mains. The machines allow, amongst other things, full audit systems, to monitor the payments being made. In addition, most machines now allow payment using a cashless system, usually a stored value card the size of a credit card. The sophistication of the equipment means that many machines have the capability for remote monitoring of their operation, either by radio or using a

Fig. 9.9 Parking voucher.

telecom line. Remote monitoring allows the operator to be warned if, for example, the machine develops a fault, is running low on tickets or has been broken into.

Parking vouchers are the most simple form of paid parking system. The voucher (Figure 9.9) has to be prepurchased, usually from a local shop. The parker scratches off the panels which relate to the time of parking and displays the card in the car's window. This can then be checked by patrol staff, to ensure that the time paid for has not been exceeded. Parking voucher design is specified in BS 6571 Part 7.[11]

9.6.3   OFF-STREET PARKING

Off-street parking can also be controlled using a traffic order. The most common form of control is to use pay and display machines. This has the advantage of being well understood by the public and relatively cheap.

Other types of parking control can also be used for off-street parking. The main methods of control are summarised in Table 9.2.

Table 9.2 Other types of control for off-street parking

System	Description	Features
Manual, pay on entry	driver pays an attendant on entry to the car park	very simple, only allows a flat fee, weak accounting of income
Manual, pay on exit	driver pays an attendant before exiting the car park	simple, only allows a flat fee unless a time-stamped ticket is issued on arrival, weak accounting of income
Automatic, pay on entry	payment at a barrier on arrival	only possible to use a flat fee, better accounting, requires either prepaid card or season ticket or cash, relatively low lane throughput (200 vph)
Automatic, pay on exit	payment at barrier on exit	only possible to use a flat fee unless linked to an entry barrier which issues a time-stamped ticket, better accounting, requires either prepaid card or season ticket or cash, relatively low lane throughput (200 vph)
Pay on foot	driver takes a time-stamped ticket on entry to the car park; on return, the driver pays on foot for parking at a pay station and then takes an exit ticket which allows exit from the car park	highly sophisticated systems allowing payment away from the entry/exit lanes, uses coins, banknotes or credit cards, separating the payment action from access barrier throughput can be increased to over 400 vph

References

1. Department of Transport (1994) *Traffic Signs Regulations and General Directions*, SI 1994/1519, HMSO, London.
2. Department of Transport (1984) *Road Traffic Regulation Act, 1984*, HMSO, London.
3. Department of Transport (1989) *Local Authorities' Traffic Order (Procedure) (England and Wales) Regulations 1989*, SI 1989/1120, HMSO, London.

4. Department of Transport (1993) *Local Authorities' Traffic Order (Procedure) (England and Wales) (Amendment) Regulations 1993,* SI 1993/1500, HMSO, London.
5. Institution of Highways & Transportation and Institution of Structural Engineers (1984) *Design Recommendations for Multi-Storey and Underground Car Parks* (2nd edn), Institution of Structural Engineers, London.
6. Association of Chief Police Officers (undated) *Secured by Design,* a Scheme for Secure Car Parks, ACPO/AA.
7. British Standards Institution (to be published) *BS 6571 Part 8: Parking Discs,* BSI, London.
8. British Standards Institution (1989) *BS 6571 Part 1: Clockwork Parking Meters,* BSI, London.
9. British Standards Institution (1989) *BS 6571 Part 2: Electronic Parking Meters,* BSI, London.
10. British Standards Institution (1989) *BS 6571 Part 3: Pay and Display Parking Meters,* BSI, London.
11. British Standards Institution (to be published) *BS 6571 Part 7: Parking Vouchers,* BSI, London.

10
Road Safety Engineering

10.1   Factors resulting in accidents

There are three factors that result in accidents:

- road and environment deficiencies
- road user errors (human factors)
- vehicle defects.

Road and environment deficiencies account on their own for only 2% of all accidents but in combination with road user errors account for slightly less than 20%. Human factors on their own account for 75% of accidents.

Typical road and environment deficiencies are those which provide misleading visual information or insufficient or unclear information to the road user. Only occasionally are accidents caused solely by bad design.

Human factors include excessive speed for the conditions, failing to give way, improperly overtaking or following too close and general misjudgement by both driver and pedestrian.

10.2   Road accident definition

The two basic types of road accident, which by definition have to involve a vehicle, are:

- personal injury
- damage only.

A personal injury accident (PIA) is an accident involving an injury. The PIA refers to the accident as the event and may involve several vehicles and several casualties (persons injured). The accident must occur in the public highway (including footways) and become known to the police within 30 days of its occurrence. The vehicle need not be moving and it need not be in collision with anything.

A casualty is a person killed or injured in an accident. Casualties are subdivided into killed, seriously injured and slightly injured. The definitions for these three subdivisions of severity are:

- Killed – a casualty who dies within 30 days of the accident but excluding confirmed suicides.
- Seriously injured – an injury for which a person is detained in hospital as an in-patient, or any of the following injuries whether or not they are detained in hospital:

fractures, concussion, internal injuries, crushing, severe cuts and lacerations, severe general shock requiring medical treatment, injuries causing death 30 or more days after the accident.

- Slightly injured – an injury of a minor character such as a sprain, bruise or cut which is not judged to be severe, or slight shock requiring road side attention. This definition includes some injuries not requiring medical treatment.

An injured casualty is recorded as serious or slightly injured by the police on the basis of information available within a short time of the accident. This generally will not include the results of a medical examination, but may be influenced by whether the casualty is hospitalised or not. There is clearly some scope for lack of uniformity of recording.

A fatal PIA is one in which there is at least one killed casualty. A serious PIA is one with no persons killed but at least one seriously injured casualty. A slight PIA is one with no persons killed or seriously injured but at least one slightly injured casualty.

Drivers have legal obligations for reporting accidents but the reporting does not necessarily have to be to the police. The police only record accidents which are reported to them. On average only 70% of accidents involving personal injury are recorded by the police. The recording of personal injury accidents involving only cyclists and no motorised vehicle is about 25%.

The TRL have determined that there are, in reality, many damage-only accidents to every accident involving personal injury.

The police are the primary source of information on road accidents. Other sources include:

- motor insurance companies
- hospital casualty (accident and emergency) units.

10.3   Police Recording

The basic accident data itself is collected by the police, being recorded in a notebook using a local standard format. In most cases this takes place as a result of attending a road accident, but there are occasions when the accident will be reported after the event at a police station. An example of a local standard format as used by the Hampshire Constabulary is shown in Figure 10.1.

There are 51 police authorities in Great Britain and the methods of handling accident data vary considerably between them. Each police force will usually have their own particular form for the accident data, which allows for completion of the STATS19 form (to be described later) and the additional information that is passed to the local authority. The additional information includes contributory factors (e.g. excessive speed) which in the police's view have contributed to the accident. In almost all police authorities, the accident data will be entered into a computer database and certain validation checks carried out. This may be done locally at individual police stations, or there may be a central unit to process the data for the whole police area. In some cases local highway authority staff may be involved.

The issue of the accuracy of the police accident data is one that is often debated between a highway authority and the respective police authority. Although it is clearly important for the data to be as accurate as possible, collecting the information is not the first priority when police officers attend the scene of an accident. The extent of reporting of accidents to police stations after the event must also be taken into account, especially as the police themselves often do not go to the site.

Some police authorities record and keep limited information on damage-only accidents which are reported to them, or which are attended. In these cases there is an additional source of data that may be available to highway authorities to assist in identifying locations for engineering remedial treatment.

 **Hampshire Constabulary**

INJURY ROAD COLLISION REPORTS

Notes for guidance for the completion of road collision reports - T1A

Shaded sections are for computer operator use only

 1 **ATTENDANT CIRCUMSTANCES**

Form Number
If only one form is used, enter 1. For additional forms number consecutively.

Collision Reference Number
To contain Divisional/Section identification letters and collision reference number. No reference to year is required.

Date
Unused boxes to the left of the day or month are to be entered as zeros; thus 9th May 1992 is coded 09 05 92.

Time
Use the 24 hour clock.

Total Number of Casualties/Vehicles
Enter total number of casualties in collision.
Enter total number of vehicles in collision.

Contributory Factors
Enter up to three factors selected from list overleaf.

Speed Limit
Enter speed limit applicable on the road. Speed limits which are temporarily in force should not be included.

Local Authority
Consult Stats 20, para. 1.10

1st Road Class and Number
Enter the class and number of road on which the collision occurred. If unclassified enter 'U'.

Marker Post Number
Enter kilometres and tenths, using leading zeros as necessary. Tenths should be entered in final box.

2nd Road Class and Number
For collisions occurring at junctions only.

Light Conditions
Codes 1- 3 apply to daylight collisions
Codes 4- 7 apply to darkness collisions.

 2 **CASUALTY DETAILS**

Each casualty should be numbered consecutively. (i.e. 1,2,3 on form 1 and 4,5,6 on form 2 etc.)

Vehicle Reference Number
To identify the vehicle occupied by a casualty prior to the collision.

Pedestrian casualty records should quote the vehicle reference number of the vehicle by which the pedestrian is first hit.

Casualty Severity
Fatal injury includes only those cases where death occurs in less than 30 days as a result of the collision. Fatal does not include death from natural causes or suicide.

Serious injury - examples include: fracture, internal injury, severe cuts and lacerations, crushing, concussion, severe general shock requiring hospital treatment, detention in hospital as an in-patient and injuries to casualties who die 30 or more days after the collision.

Slight Injury - examples include: sprains, bruises, cuts judged not to be severe, slight shock requiring road side attention.

Car Passenger
In case of drivers, tick '0 - not car passenger.'

DoT Special Projects Local Special Projects
Use only when directed.

School Pupil
Tick box 1 *only* if the casualty is a school pupil on the journey to or from school. All other casualties, including school age children *not* going to or from school, tick box 0.

School pupils are children aged between 5 and 16 years inclusive.

NB
First row or column refers to the first casualty/vehicle on form
Second row or column refers to second casualty/vehicle on form
Third row or column refers to third casualty/vehicle on form

 3 **VEHICLE DETAILS**

Vehicle Reference Number
Each vehicle is to be numbered consecutively (i.e. 1,2,3 on form 1 and 4,5,6 on form 2 etc.)

Type of Vehicle
The following types of vehicle are defined -
Box 8 - Taxi - any licensed hackney vehicle carrying appropriate local authority plates; but does not include private hire vehicles.
Box 11 - Bus or Coach - any passenger carrying vehicle (PCV) equipped to carry 17 or more persons.
Boxes 12 and 13 - goods vehicles are classified here by maximum gross weight; either up to and including 3.5 tonnes or over 3.5 tonnes.

Postcode
Give first half of postcode of driver's home address, if known, e.g. if code is SO22 5DB enter SO22. Part code, e.g. SP is acceptable if remainder is not known. Do not guess if code is not known.

Other Vehicle Hit
Enter the vehicle reference number of the first other vehicle (if any) with which the vehicle being coded collided.

DoT Special Projects Local Special Projects
Use only when directed

Overshoot/Restart
Indicate whether vehicle stopped and restarted at junction (restart) or failed to stop at junction (overshoot).

Part(s) Damaged
Up to three codes may be ticked in this section for each vehicle.

Breath Test
In cases where the breathalyser procedure cannot in law be applied, e.g. in the case of non-motor vehicles. tick '0 - Not applicable.'

Where a negative breath test has been given T28A *must* also be submitted.

Fig. 10.1 Example of an injury road collision report.

10.4 National Accident Reporting System and STATS19

The National Accident Reporting System defines and records accidents as only personal injury accidents on the public highway (including footway) in which a vehicle is involved and which became known to the police within 30 days of its occurrence.

The vehicle does not need to be moving and it need not be in collision with anything. A PIA could then include cases of accidents involving people boarding or alighting from vehicles or of injuries when vehicles brake.

The designation of the forms used for the National Accident Reporting System is STATS19. STATS19 has a fixed format and consists of three records which must be completed.

The Accident Record Attendant Circumstances describe the physical nature of the location of the accident. The Vehicle Record contains information on the vehicle involved (but not the registration number) and the driver, with a separate record for each vehicle. The Casualty Record includes more variables dealing with pedestrian casualties than vehicle driver and occupant casualties, to deal with such information as pedestrian location and movement, with a separate record for each casualty involved. These forms are included as Figures 10.2–4.

The only personal information shown in STATS19 is age and gender of casualties and drivers and thus it is not affected by the Data Protection Act. There are linkages between the three records, for example, to identify in which particular vehicle each vehicle casualty was travelling.

The publication *STATS20: Instructions for the Completion of Road Accident Report Form STATS19* is the manual which defines each of the variables in detail, as well as showing how the STATS19 form should be completed.

The STATS19 data is passed by the police through local highway authorities to the Department of Transport on a monthly basis, when it is checked and validated, before being added to the national database. Every three months, a bulletin is published, showing brief provisional quarterly accident and casualty figures, compared with the same period for the preceding year. An annual DoT report[1] has been published for a number of years, showing comprehensive analyses of the accident variables in terms of accidents and casualties for the latest year, together with the analysis of trends.

10.5 Data transfer to the highway authority

In addition to the fixed format STATS19 data, local highway authorities need to collect other accident information. The exact nature and extent of this additional information will depend upon the arrangements that have been agreed with the respective police authority(s). However, for road safety engineering, it is essential that there is a description and location of the accident in plain language, in addition to the STATS19 information. In many circumstances, the description is vital in piecing together the events leading up to the accident, particularly where there is conflicting evidence from the STATS19 variables.

In rural areas, it is not always possible to determine the location from the grid reference alone and, furthermore, the location allows the original grid coordinates to be checked. Other information that can be used in some road safety engineering studies, is the school attended in the case of child pedestrian casualties. In many cases, the police will also include what is called a 'contributory or causation factor', which is a subjective assessment as to the primary cause of an accident, with often similar factors allocated to each casualty and vehicle. The Hampshire Constabulary's Contributory Factors list is shown as Figure 10.5; each constabulary has its own list of factors.

Department of Transport

Accident Record Attendant Circumstances

Stats 19 (Rev 2/91)

1.1 Record Type
1 New accident record
5 Amended accident record
`[1]` 1 2

1.2 Police Force
`[ ]` 3 4

1.3 Accident Ref No
`[    ]` 5 6 7 8 9 10 11

1.4 Severity of Accident
1 Fatal 2 Serious 3 Slight
`[ ]` 12

1.5 Number of Vehicles
`[  ]` 13 14 15

1.6 Number of Casualty Records
`[  ]` 16 17 18

1.7 Date
Day `[  ]` 19 20
Month `[  ]` 21 22
Year `[  ]` 23 24

1.8 Day of Week
1 Sunday 2 Monday
3 Tuesday 4 Wednesday
5 Thursday 6 Friday
7 Saturday
`[ ]` 25

1.9 Time
Hrs `[  ]` 26 27
Mins `[  ]` 28 29
24 hour

1.10 Local Authority
`[    ]` 30 31 32

1.11 Location
10 digit reference No
Easting `[     ]` 33 34 35 36 37
Northing `[     ]` 38 39 40 41 42
`[ ]` 43

1.12 1st Road Class
1 Motorway
2 A (M)
3 A
4 B
5 C
6 Unclassified
7 Local
8 Authority
9 Use Only
`[ ]`

1.13 1st Road Number
`[    ]` 44 45 46 47

1.14 Carriageway Type or Markings
1 Roundabout (on circular highway)
2 One way street
3 Dual carriageway - 2 lanes
4 Dual carriageway - 3 or more lanes
5 Single carriageway - single track road
6 Single carriageway - 2 lanes (one each direction)
7 Single carriageway - 3 lanes (two way capacity)
8 Single carriageway - 4 or more lanes (two way capacity)
9 Unknown
`[ ]` 48

1.15 Speed Limit
mph
`[0 ]` 49 50 51

1.16 Junction Detail
0 Not at or within 20 metres of junction
1 Roundabout
2 Mini roundabout
3 'T' or staggered junction
4 'Y' junction
5 Slip road
6 Crossroads
7 Multiple junction
8 Using private drive or entrance
9 Other junction
`[0 ]` 52 53

Junction Accidents Only

1.17 Junction Control
1 Authorised person
2 Automatic traffic signal
3 Stop sign
4 Give way sign or markings
5 Uncontrolled
`[ ]` 54

1.18 2nd Road Class
1 Motorway
2 A (M)
3 A
4 B
5 C
6 Unclassified
7 Local
8 Authority
9 Use Only
`[ ]` 55

1.19 2nd Road Number
`[    ]` 56 57 58 59

1.20 Pedestrian Crossing Facilities
0 No crossing facilities within 50 metres
1 Zebra
2 Zebra crossing controlled by school crossing patrol
3 Zebra crossing controlled by other authorised person
4 Pelican
5 Other light controlled crossing
6 Other sites controlled by school crossing patrol
7 Other sites controlled by other authorised person
8 Central refuge - no other controls
9 Footbridge or subway
`[0]` 60 61

1.21 Light Conditions
DAYLIGHT
1 Street lights 7 metres or more high
2 Street lights under 7 metres high
3 No street lighting
4 Daylight street lighting unknown
DARKNESS
5 Street lights 7 metres or more high (lit)
6 Street lights under 7 metres high (lit)
7 No street lighting
8 Street lights unlit
9 Darkness street lighting unknown
`[ ]` 62

1.22 Weather
1 Fine (without high winds)
2 Raining (without high winds)
3 Snowing (without high winds)
4 Fine with high winds
5 Raining with high winds
6 Snowing with high winds
7 Fog (or mist if hazard)
8 Other
9 Unknown
`[ ]` 63

1.23 Road Surface Condition
1 Dry
2 Wet/Damp
3 Snow
4 Frost/Ice
5 Flood (surface water over 3cms (1 inch) deep)
`[ ]` 64

1.24 Special Conditions at Site
0 None
1 Automatic Traffic Signal-out
2 Automatic Traffic Signal partially defective
3 Permanent road signing defective or obscured
4 Road works present
5 Road surface defective
`[ ]` 65

1.25 Carriageway Hazards
0 None
1 Dislodged vehicle load in carriageway
2 Other object in carriageway
3 Involvement with previous accident
4 Dog in carriageway
5 Other animal in carriageway
`[ ]` 66

1.26 Overtaking Manoeuvre Patterns
No longer required by the Department of Transport
`[ ]` 67

1.27 DTp Special Projects
`[    ]` 68 69 70 71

Fig. 10.2 Accident record attendant circumstances.

Vehicle Record

2.1 Record Type
1 New vehicle record
5 Amended vehicle record
[2] (1 2)

2.2 Police Force
[ ] (3 4)

2.3 Accident Ref No
[ ] (5 6 7 8 9 10 11)

2.4 Vehicle Ref No
[ ] (12 13 14)

2.5 Type of Vehicle
[ ] (15 16)
01 Pedal cycle
02 Moped
03 Motor scooter
04 Motor cycle
05 Combination
06 Invalid Tricycle
07 Other three wheeled car
08 Taxi
09 Car (four wheeled)
10 Minibus/Motor caravan
11 PSV
12 Goods not over 1 1/2 tons UW (1.52 tonnes)
13 Goods over 1 1/2 tons UW (1.52 tonnes)
14 Other motor vehicle
15 Other non motor vehicle

2.6 Towing and Articulation
[ ] (17)
0 No tow/articulation   1 Articulated vehicle
2 Double/multiple trailer 3 Caravan
4 Single trailer   5 Other tow

2.7 Manoeuvres
[ ] (18 19)
01 Reversing
02 Parked
03 Waiting to go ahead but held up
04 Stopping
05 Starting
06 U turn
07 Turning left   08 Waiting to turn left
09 Turning right   10 Waiting to turn right
11 Changing lane to left
12 Changing lane to right
13 Overtaking moving vehicle on its offside
14 Overtaking stationary vehicle on its offside
15 Overtaking on nearside
16 Going ahead left hand bend
17 Going ahead right hand bend
18 Going ahead other

2.8 Vehicle Movement Compass Point
[ ] [ ] (20 21) From To
1 N   2 NE   3 E
4 SE 5 S   6 SW
7 W   8 NW
or [0][0] Parked - not at kerb
[0] Parked - at kerb

2.9 Vehicle Location at time of Accident
[ ] (22 23)
01 Leaving the main road
02 Entering the main road
03 On main road
04 On minor road
05 On service road
06 On lay by or hard shoulder
07 Entering lay by or hard shoulder
08 Leaving lay-by or hard shoulder
09 On a cycleway
10 Not on carriageway

2.10 Junction Location of Vehicle at First Impact
[ ] (24)
0 Not at junction (or within 20 metres/22 yards)
1 Vehicle approaching junction/vehicle parked at junction approach
2 Vehicle in middle of junction
3 Vehicle cleared junction/vehicle parked at junction exit
4 Did not impact

2.11 Skidding and Overturning
[ ] (25)
0 No skidding, jackknifing or overturning
1 Skidded   2 Skidded and overturned
3 Jackknifed   4 Jackknifed and overturned
5 Overturned

2.12 Hit Object In Carriageway
[ ] (26 27)
00 None
01 Previous accident
02 Road works
03 Parked vehicle - lit
04 Parked vehicle - unlit   05 Bridge (roof)
06 Bridge (side)   07 Bollard/refuge
08 Open door of vehicle
09 Central island or roundabout
10 Kerb   11 Other object

2.13 Vehicle Leaving Carriageway
[ ] (28)
0 Did not leave carriageway
1 Left carriageway nearside
2 Left carriageway nearside and rebounded
3 Left carriageway straight ahead at junction
4 Left carriageway offside onto central reservation
5 Left carriageway offside onto central reservation and rebounded
6 Left carriageway offside crossed central reservation
7 Left carriageway offside
8 Left carriageway offside and rebounded

2.14 Hit Object Off Carriageway
[ ] (29 30)
00 None   01 Road sign/Traffic signal
02 Lamp post
03 Telegraph pole/Electricity pole
04 Tree   05 Bus stop/Bus shelter
06 Central crash barrier
07 Nearside or offside crash barrier
08 Submerged in water (completely)
09 Entered ditch
10 Other permanent object

2.15 Vehicle Prefix/Suffix Letter
[ ] (31)
Prefix/Suffix letter or one of the following codes -
0 More than twenty years old (at unit of year)
1 Unknown/cherished number/not applicable
2 Foreign/diplomatic
3 Military
4 Trade plates

2.16 First Point of Impact
[ ] (32)
0 Did not impact
1 Front   2 Back
3 Offside   4 Nearside

2.17 Other Vehicle Hit (Veh Ref No)
[ ] (33 34 35)

2.18 Part(s) Damaged
[ ] (36 37 38)
0 None   1 Front
2 Back   3 Offside
4 Nearside   5 Roof
6 Underside   7 all four sides

2.19 No of Axles
[ ] (39)
No longer required by the Department of Transport

2.20 Maximum Permissible Gross Weight
[ ] (40 41)
Metric tonnes (Goods vehicle only)

2.21 Sex of Driver
[ ] (42)
1 Male   2 Female
3 Not traced

2.22 Age of Driver
[ ] (43 44)
(Years estimated if necessary)

2.23 Breath Test
[ ] (45)
0 Not applicable   1 Positive
2 Negative   3 Not requested
4 Failed to provide
5 Driver not contacted at time

2.24 Hit and Run
[ ] (46)
0 Other   1 Hit and run
2 Non stop vehicle not hit

2.25 DTp Special Projects
[ ] (47 48 49 50)

Fig. 10.3 Vehicle record.

3.1 Record Type `1 2` [3]
1 New casualty record
5 Amended casualty record

3.2 Police Force `3 4`

3.3 Accident Ref No `5 6 7 8 9 10 11`

3.4 Vehicle Ref No `12 13 14`

3.5 Casualty Ref No `15 16 17`

3.6 Casualty Class `18`
1 Driver or Rider
2 Vehicle or pillion passenger
3 Pedestrian

3.7 Sex of Casualty `19`
1 Male
2 Female

3.8 Age of Casualty `20 21`
(Years estimated if necessary)

3.9 Severity of Casualty `22`
1 Fatal
2 Serious
3 Slight

3.10 Pedestrian Location `23 24`
00 Not pedestrian
01 In carriageway crossing on pedestrian crossing
02 In carriageway crossing within zig-zag lines approach to the crossing
03 In carriageway crossing within zig zag lines exit the crossing
04 In carriageway crossing elsewhere within 50 metres of pedestrian crossing
05 In carriageway crossing elsewhere
06 On footway or verge
07 On refuge or central island or reservation
08 In centre of carriageway not on refuge or central island
09 In carriageway not crossing
10 Unknown

3.11 Pedestrian Movement `25`
0 Not pedestrian
1 Crossing from drivers nearside
2 Crossing from drivers nearside - masked by parked or stationary vehicle
3 Crossing from drivers offside
4 Crossing from drivers offside - masked by parked or stationary vehicle
5 In carriageway stationary - not crossing (standing or playing)
6 In carriageway stationary - not crossing (standing or playing) - masked by parked or stationary vehicle
7 Walking along in carriageway facing traffic
8 Walking along in carriageway back to traffic
9 Unknown

3.12 Pedestrian Direction `26`
Compass point bound
1 N
2 NE
3 E
4 SE
5 S
6 SW
7 W
8 NW
or 0 - Pedestrian - standing still

3.13 School Pupil Casualty `27`
0 Not a school pupil
1 Pupil on journey to/from school
2 Pupil NOT on journey to/from school

3.14 Seat Belt Usage `28`
0 Not car or van
1 Safety belt in use
2 Safety belt fitted - not in use
3 Safety belt not fitted
4 Child safety belt/harness fitted - in use
5 Child safety belt/harness fitted - not in use
6 Child safety belt/harness not fitted
7 Unknown

3.15 Car Passenger `29`
0 Not a car passenger
1 Front seat car passenger
2 Rear seat car passenger

3.16 PSV Passenger `30`
0 Not a PSV passenger
1 Boarding
2 Alighting
3 Standing passenger
4 Seated passenger

3.17 DTp Special Projects `31 32 33 34`

Fig. 10.4 Casualty record.

A: Collisions due to actions of drivers/riders

    1 Tired or asleep
    2 Illness
    3 Drunk or drugged
    4 Speed too great for prevailing conditions
    5 Failing to keep to nearside
    6 Overtaking improperly on nearside
    7 Overtaking improperly on offside
    8 Failing to stop at pedestrian crossing
    9 Turning round carelessly
    10 Reversing carelessly
    11 Failing to comply with traffic sign
       (other than double white line or traffic lights)
    12 Failing to comply with double white lines
    13 Starting from nearside carelessly
    14 Starting from offside carelessly
    15 Changing traffic lanes carelessly
    16 Cyclist riding with head down
    17 Cyclists more than two abreast
    18 Turning left carelessly
    19 Turning right carelessly
    20 Opening doors carelessly
    21 Crossing road junction carelessly
    22 Cyclist holding another vehicle
    23 Not in use
    24 Misjudging clearance
    63 Failing to comply with traffic lights
    64 Incorrect use of vehicle lighting

B: Collisions due to actions of pedestrians

    26 Crossing road masked by vehicle
    27 Walking or standing in road
    28 Playing in road
    29 Stepping or running into road carelessly
    30 Physical defects or illness
    31 Drunk or drugged
    32 Holding onto vehicle

C: Vehicle lighting

    33 Dazzle by other vehicle's lights
    34 Inadequate rear lights
    35 Inadequate front lights
    36 Not in use

D: Collisions due to actions of passengers

    37 Carelessly boarding or alighting from bus
    38 Falling inside or from vehicle
    39 Opening door carelessly
    40 Negligence by conductor of bus

E: Collisions due to actions of animals

    41 Dog in carriageway
    42 Other animal in carriageway

F: Collisions due to obstructions

    43 Stationary vehicles dangerously placed
    44 Other obstructions

G: Collisions due to defective vehicles

    45 Defective brakes
    46 Defective tyres or wheels
    47 Defective steering
    48 Unattended vehicle running away
    49 Insecure load
    50 Other defects

H: Collisions due to road conditions

    51 Pot hole
    52 Defective manhole cover
    53 Other road surface conditions
    54 Road works in progress
    55 Slippery road surface (not weather)
    65 Flood

I: Collisions due to weather conditions

    56 Fog or mist
    57 Ice, frost or snow
    58 Strong wind
    59 Heavy rain
    60 Glaring sun

J: Collisions due to other factors

    61

K: Collisions due to unknown factors

    62

    USE 61 OR 62 ONLY IF ALL OTHER FACTORS
    ARE INAPPROPRIATE

Fig. 10.5 Contributory factors.

On receipt of the data each month from the police, a number of automatic computer validation checks will be carried out. In many cases, local highway authorities will check the grid coordinates manually, with perhaps some other basic manual checking being carried out.

There are some common accident systems that are used by a number of highway authorities, but in many cases there will be a specific purpose-designed system.

The source of accident data for the users of this data, be they local authority officers or consultants, is normally the highway authority. Sometimes the information can only be obtained by applying to the police. In London, the London Research Centre holds the data on behalf of the boroughs. Joint data units carry out a similar function in other metropolitan areas.

10.6   Other data to be collected for safety work

Road safety engineers will normally wish to collect several other items of information. These include:

- up-to-date 1:1250 (or 1:2500) mapping
- classified traffic counts or turning movements
- annual average traffic volumes
- pedestrian flows
- traffic speeds
- population
- street inventory (traffic signals, pedestrian crossings, banned turns, one-way streets, pedestrian guardrail)
- local traffic attractors, e.g. shops, offices, hospitals, schools.

10.7   Presenting accident data for analysis

10.7.1   REGRESSION TO THE MEAN

It is a truism to say that accidents are rare events subject to random variation. Random variation will have a biasing effect if combined with a tendency to select sites for treatment (i.e., improvement) on the basis of accident records in the recent past. This is because a selection process based on recent past accident records tends to produce sites for treatment which happen to be at a peak in their fluctuation of accident frequency over time. We should expect such sites to experience reductions in accident frequency in subsequent years, regardless of whether remedial measures are implemented. This effect is commonly known as 'regression to the mean'.

It is desirable therefore to present accident data for analysis over a long period – at least 3 years – and to compare with control sites of a similar nature in the same area. Highway authorities find it difficult to conduct control programmes because of the requirements of the Road Traffic Act.

10.7.2   GRAPHICAL

Presenting accident data on a map background provides an immediate identification of the locations, i.e. junctions and road sections, where accidents have occurred. Digitised mapping is now commonly used.

It is useful to record the accidents separately as slight, serious or fatal so that the graphical presentation can also provide a measure of severity by location.

10.7.3   STICK DIAGRAMS

When a particular location is to be studied in more detail, it is useful to present the information as a stick diagram, as in Figure 10.6. The vehicle manoeuvre for each accident often has to be added manually, using symbols which vary by authority.

10.7.4   HISTOGRAMS (CONTINGENCY TABLES)

Histograms are useful when studying a location to identify special accident features, e.g.:

- vehicle category
- time of day
- month of year
- day of week
- dry, wet or icy road surface
- daylight or darkness (lit or unlit)
- age of casualty
- type of casualty.

The histogram can include national averages for comparison purposes as in Figure 10.7.

10.7.5   TYPE OF ACCIDENT

A series of graphical presentations can provide information in more detail for the same area. The detail could include vehicle type and vulnerable road user (pedestrian, cyclist), by age or by a particular type of accident, such as overtaking.

10.8   Summarising data

Highway authorities will want to summarise accident data for the whole of their area, to compare with national or local road safety targets. The national road safety target set in 1987 was:

> A one third reduction in all road casualties by the year 2000 with respect to the average level of casualties for the years 1981–85.

Achievement against this target is shown in Table 10.1. Although there has already been more than a one-third reduction in killed and seriously injured casualties, the overall target is not going to be met. This is primarily due to the growth in slightly injured casualties, although there is a suspicion that part of the growth is due to a decline in the proportion of unreported slight injuries.

Road accident statistics by region and county are available from the DoT.[2] Table 10.2 provides some examples of reduction in casualties between the 1981–85 baseline and 1995.

The national accident rates per 100 million vehicle kilometres have fallen in the period to 1995 by 35% on built-up roads, 27% on non-built-up roads and 23% on motorways.

The percentage change in accidents by type of road user is shown in Table 10.3 with the current rates for killed and seriously injured (KSI) casualties by road user shown in Table 10.4.

There is a debate now as to the nature of the next national road safety target. It appears likely that there will be a disaggregation of the current target associated with key areas for action, such as vulnerable road users.

MVA REFERENCE NO:	2	8	47	111	7	74			
DAY	THURSDAY	TUESDAY	THURSDAY	FRIDAY	SUNDAY	SATURDAY			
DATE	11/08/83	13/09/88	21/12/89	08/01/93	04/09/88	29/12/90			
TIME	1250	1420	1235	1745	1630	1740			
DARK/L				DARK					
WET/D			WET	WET		WET			
SEVERITY	PED/SLIGHT	PED/SLIGHT	PED/SLIGHT	PED/SERIOUS	VEH/SLIGHT	VEH/SLIGHT			
CONFLICT									
AGE OF CASUALTY	13								
CONTRIB F	D	D	OC		U	I			
EASTING	517950	517940	517950	517950	517950	517950			
NORTHING	104390	104400	104400	104400	105200	104390			

Fig. 10.6 Accident analysis used by West Sussex County Council Urban Safety Management.

Year ending	No. of accidents		
30.4.83	3	FAT	0
		SER	2
		SLT	1
30.4.84	6	FAT	0
		SER	3
		SLT	3
30.4.85	3	FAT	0
		SER	2
		SLT	1
Total	12	FAT	0
		SER	7
		SLT	5

Accident severity by year

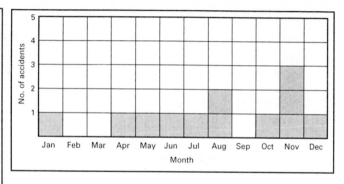

Vehicle category	No.	%	Reg. %
Car/Taxi	13	50	63.6
Goods	11	42.3	8.8
PSV	0	0	3.1
Motor cycle	2	7.7	13.7
Pedal cycle	0	0	6.3
Other	0	0	4.5
Total	26	100%	100%

Vehicle category

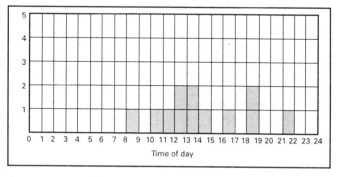

Casualty class	No.	%	Reg. %
Driver			54.5
Passenger	20 — 100		26.7
Pedestrian			18.8
Total	20	100%	100%

Surface	No.	%	Reg. %
Dry	10	83.3	62
Wet	2	16.7	33.9
Snow	–	–	4.1

Road surface condition

	No.	%	Reg. %
Skid	1	8.3	27
No skid	11	91.7	73

Skid/no skid

Age	No.	%	Reg. %
0 – 15	0	0	15.5
16 – 19	3	15	22.4
20 – 29	10	50	24.3
30 – 59	6	30	27.3
60+	1	5	10.5

Casualty age

	No.	%	Reg. %
Daylight	12	100	67
Darkness		0	33

Daylight/darkness

Fig. 10.7 Accident histograms.

Table 10.1 Casualties in UK (excepting Northern Ireland)

Casualty	1981/85 baseline	1995	Change	Change (%)
Killed	5 598	3 621	−1 977	−35
Seriously injured	74 533	45 523	−29 010	−39
Slightly injured	241 787	261 362	+19 575	+8
All casualties	321 918	310 506	−11 412	−4

Table 10.2 Selected regional data – percentage reduction in casualties 1981/5–1995

	Killed and seriously injured	All casualties
Greater London	−20.9	−16.9
Kent	−41.3	−11.9
Hampshire	−49.9	−3.3
Oxfordshire	−63.0	−10.0
South Yorkshire	−43.5	0.2
Warwickshire	−37.2	8.6
West Midlands	−40.4	−2.6
West Sussex	−34.0	−4.8
England	−38.2	−2.5
Wales	−44.7	3.9
Scotland	−40.0	−18.2
Great Britain	−38.7	−3.5
Northern Ireland	−29.0	42.9
United Kingdom	−38.4	−2.4

Table 10.3 All casualties by road user

User	1981/85 baseline	1995	Change	Change (%)
Pedestrians	61 742	47 029	−14 713	−24
Pedal cyclists	28 391	24 913	−3 478	−12
Two-wheeled motor vehicles (TWMV)	65 193	23 480	−41 713	−64
Car users	143 944	193 992	+50 048	+35
Bus and coach	10 182	9 269	−913	−9
Goods vehicles	11 196	10 526	−670	−6

Table 10.4 Passenger casualty rates (KSI) by mode of travel

Mode	Rate per 100 million		
	Passenger kilometres	Passenger journeys	Passenger hours
Car	4.5	55	190
Van	2.4	30	75
Two-wheeled motor vehicle	150.0	1600	4500
Pedal cycle	85.0	230	1200
Foot	55.0	55	225
Bus or coach	1.6	13	40

10.9   Data presentation and ranking

10.9.1   FOUR APPROACHES

The basis of the UK structured system for accident investigation and prevention is the use of four main approaches:

- Single sites – treatment of specific sites or short lengths of road where clusters of accidents have occurred.
- Mass action plans – application of a known remedy to locations having common accident factors.
- Route action plans – application of remedies to a road having an above average accident rate for that type of road and traffic volume.
- Area action plans – remedial measures over an area with an above average accident rate for that type/size of area or population.[3]

The method of presentation of results, ranking and numerical/statistical analysis depends on the approach.

10.9.2   ACCIDENT TOTALS AND RATES

Sites, such as junctions, can be listed in order of numbers of accidents. Road sections can be listed in rate order as numbers of accidents per 100 million vehicle kilometres. Areas can be listed in rate order as numbers of accidents per square kilometre or population. Listing accidents in rate order instead of number order can identify sites of high risk for lower numbers of users.

10.9.3   POTENTIAL ACCIDENT REDUCTION

Another way of ranking for treatment is to present the accident data for each site/road section/area in terms of potential accident reduction (PAR). This can be used in combination with totals and rates for choosing top sites for analysis.

10.9.4   TYPE OF ACCIDENT OR USER

For mass action plans the engineer will be concentrating on ranking for a particular type of accident or road user. The type of accident could be skidding, dark (night-time) hours or red-light running – for action plans associated with skid-resistance surfacing, improved lighting or introduction of red-light cameras. (Red-light running is the disobeyance by drivers of the red aspect at traffic signals.)

The type of road user could be pedestrian or cyclist, with action plans for introducing pedestrian and cyclist safety improvements.

10.9.5   ACCIDENTS PER SQUARE KILOMETRE OR POPULATION

For area action plans the engineer will be ranking areas by identifying accidents per square kilometre or resident population statistic. This could be used for ranking areas of accidents of a similar nature (e.g. involving vulnerable road users) for traffic calming treatment.

10.9.6   COMPARISON WITH AVERAGES

National road accident statistics, obtained from the national database, are available in various government documents. An example extraction is shown as Table 10.5.[4]

Regional and county averages are published by the highway authorities, the London Research Centre and joint data units. These averages are useful for providing a comparison against accident rates on roads in the area being analysed.

Table 10.5 Personal injury accident rates by road type and year per 100 million vehicle kilometres

Road type	Year (19..)											
	83	84	85	86	87	88	89	90	91	92	93	94
Motorways	12	12	12	13	11	11	11	11	10	11	11	11
Built-up roads (speed limit of 40 mph or less)												
A roads	130	133	128	124	115	112	108	106	96	96	95	95
Other roads	136	138	129	121	111	110	99	100	92	92	87	85
Non-built-up roads (speed limit exceeding 40 mph but excl. motorways)												
A roads	38	37	37	37	34	35	34	33	29	28	28	28
Other roads	50	49	47	47	42	44	44	44	41	41	44	44
All roads	83	82	78	75	67	67	63	62	57	56	55	55

10.10   Statistical analysis

10.10.1   SIGNIFICANCE LEVELS

A significance level for some accident statistic of $x\%$ implies that there is only an $x\%$ chance of that statistic occurring as a result of random fluctuations. We can refer to a significance level of $x\%$ or confidence level of $(100 - x)\%$ in the following terms:

Significance level	Confidence level	Significance
1%	99%	highly significant
5%	95%	significant
10%	90%	fairly significant
>20%	<80%	not significant

10.10.2   POISSON TEST

This is a useful test for calculating the probability of a particular number of accidents occurring at a location in a given year when the long-term average of numbers of accidents is known. For example, if the average annual number of accidents at a location is two and then in 1 year five accidents occur, this can be shown using Poisson tables to be significant. This is because the probability of five or more accidents occurring due to random fluctuations is only 5% and so there is a significant (95%) chance that a real increase in accidents has occurred.

10.10.3   CHI-SQUARED TEST

This is a test for determining whether the number of accidents of a certain type at a particular site are significantly different to the number at similar (control) sites. It is also used for determining whether the number of accidents at a site after remedial measures have been carried out has changed significantly with reference to similar control sites over the same period. Consider the following table:

	Numbers of accidents	
Site	Before measures	After measures
Treated site	b	a
Control site	B	A

If χ is greater than 4 then the treated site is significantly different at the 5% significance level:

$$\chi = \frac{(bA - aB)^2 (a + b + A + B)}{(b + a)(B + A)(b + B)(a + A)}$$

10.11   Problem identification

10.11.1   LOOKING FOR COMMON PATTERNS

The road safety engineer will make use of the accident data presented to him to identify common patterns. Typical patterns would be:

- above average night-time accidents associated with poor or nonexistent lighting;
- skidding accidents associated with poor surfacing or drainage;
- pedestrian accidents associated with inadequate crossings or excessive vehicle speed.

Problem identification is not a mechanical exercise and so has to be learnt from practical experience. It is often necessary to use several different methods of presentation and to refer to the traffic data and street inventory.

The engineer would normally supplement this analysis of common patterns to identify problems by one or more of the following:

- on-site observations
- conflict studies
- location sampling.

10.11.2   ON-SITE OBSERVATIONS

These involve driving, walking and observing the location over an extended period of time and possibly at different times of the day and in different weather and lighting conditions.

Photographs taken at driver eye height or pedestrian eye height provide a record of visual perception.

The site visit can, amongst other matters, also identify:

- poor sight lines (and intervisibility between traffic flows) caused perhaps by street furniture;
- confusing situation through poor road markings or inadequate maintenance of signs;
- forward visibility impaired by foliage or parked vehicles.

10.11.3   CONFLICT STUDIES

This involves observations of vehicle movements at specific locations in order to assess the frequency and type of near-accident situations.

10.11.4   LOCATION SAMPLING

Location sampling is useful in developing mass action plans. It involves the grouping of accident data for sites with similar physical or accident features to provide sufficient data for the assessment of contributory factors, such as excessive speed.

10.11.5   RECENT RESEARCH

Some recent research undertaken by the AA Foundation for Road Safety Research with Cambridgeshire County Council has identified some common aspects of accidents on rural roads that relate to road environment and engineering issues:

1. The incidence of accidents is not a simple function of traffic volume or of flow to capacity ratio.
2. A carriageway widening right-turn facility (ghost island) at T junctions reduces the number of stacking-type accidents in the approach lane remote to the minor road, although they can cause an increase in accidents for vehicles turning right out of the minor road. An untested remedy could be to change the road markings to provide a substandard 1 m right turn facility while retaining the extra carriageway width.
3. Accidents at T junctions with busy private access are a concern but the highway authority has few powers to treat them. Selective carriageway widening at these T junctions is likely to have the greatest reduction effect.
4. Accidents at T junctions involving a high proportion of dry-weather skidding imply excessive speed.
5. Skidding accidents on wet roads are highest on roads with a poor level of skidding resistance.
6. The probability of an overtaking-type accident increases if the forward visibility is considerably greater than the design standard. These accidents are more likely to occur if sight distance is substandard – (less than 215 m) or very good (more than 580 m).

TRL has undertaken considerable research on road safety and some guidance on aspects of the safety of link and junction arrangements can be obtained from that research. The London Accident Analysis Unit have also produced reports on topics covering accidents specific to the elderly, children, parked vehicles, minibuses, taxis, pedestrians, public service vehicles, pedal cyclists, signal-controlled junctions, younger drivers and vehicle speed.

10.12   Institution of Highways and Transportation accident investigation procedures

Figure 10.8 shows the components of these procedures to be:[5]

- data provision (data collection);
- problem site and situation identification (presenting accident data for analysis, ranking and numerical and statistical analysis);
- diagnosis of problem (problem identification);
- selection of site for treatment.

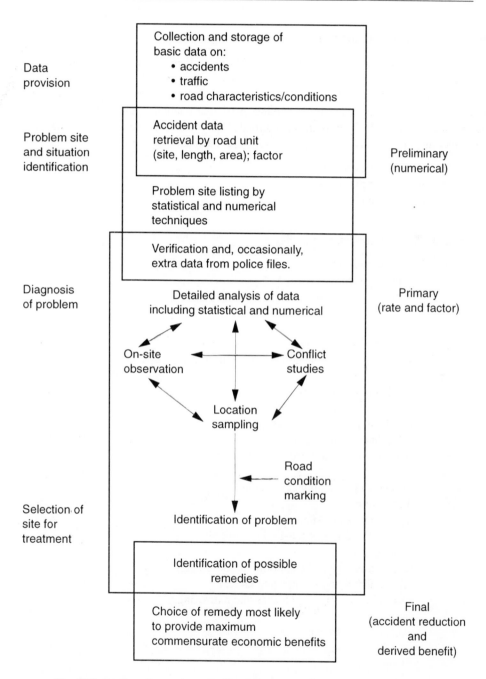

Procedure areas	Tasks	Ranking stages
Data provision	Collection and storage of basic data on: • accidents • traffic • road characteristics/conditions	
Problem site and situation identification	Accident data retrieval by road unit (site, length, area); factor Problem site listing by statistical and numerical techniques	Preliminary (numerical)
Diagnosis of problem	Verification and, occasionally, extra data from police files. Detailed analysis of data including statistical and numerical On-site observation — Conflict studies Location sampling Road condition marking Identification of problem	Primary (rate and factor)
Selection of site for treatment	Identification of possible remedies Choice of remedy most likely to provide maximum commensurate economic benefits	Final (accident reduction and derived benefit)

Fig. 10.8 Outline of procedures for identification, diagnosis and selection of sites.

The fourth procedure area, selection of site for treatment, involves identification of possible remedies and choice of remedy with assessment of accident benefit. Further guidance is available from the Royal Society for Prevention of Accidents[6] and the European Transport Safety Council.[7]

10.13   Designing road safety engineering measures

Once ranking of sites for treatment has been undertaken, then engineering measures to remedy the accident problem will be designed and accident savings assessed. Engineering measures will be prepared not only to remedy accident problems but also to improve perception of safety, for example to overcome the deterrent to walking, particularly amongst the young, the less mobile and the elderly, caused by heavily trafficked roads with inadequate crossings for pedestrians.

The measures to be considered depend on the approach adopted, such as:

- single site treatment
- mass action plans
- route action plans
- area action plans.

Typical examples of measures are:

- Single site treatment
  improved signing
  carriageway markings
  road surface treatment
  lighting improvements
  alterations to alignment, kerbs and islands
  introduction of signal control or miniroundabouts.
- Mass action plan
  lighting improvements
  antiskid surfacing
  speed enforcement cameras
  red-light running cameras
  pelicans, zebras and pedestrian phases at signals.
- Route action plan
  carriageway widening at junctions
  speed limits
  speed control measures
  side road closures or left-in, left-out only
  cycle routes.
- Area action plan – traffic calming measures, e.g.:
  vertical deflection (humps and tables)
  horizontal deflection (chicanes and narrowings)
  miniroundabouts
  road entry treatment
  road closures and banned turns
  20 mph zones.

Figures 10.9 and 10.10 show before and after views of road junctions improved as accident prevention measures.

Fig. 10.9 Accident prevention scheme at a rural crossroads (courtesy Hampshire County Council).

Fig. 10.10 Accident prevention scheme in a village centre (courtesy Hampshire County Council).

10.14   Accident savings

The procedure for assessing accident savings is pragmatic; there is no text-book definitive formula for assessment. General guidelines are available from the latest research and studies undertaken either by or on behalf of the IHT, TRL and local highway authorities. The TRL have identified that achievement of a 1 mph reduction in average vehicle speeds can reduce accidents by as much as 5%. The IHT have suggested the following guidelines for percentage accident reductions:

single site action	33%
route action	15%
redistribution and general treatment	10%
20 mph zones	50–75%

Reductions due to redistribution and general treatment require further explanation. Accident reduction can be achieved by measures designed to redistribute traffic within an area by actively discouraging the use of certain routes, which may not have particular accident problems but will benefit from a reduction in vehicle flow.

A reduction can also be achieved by general treatment of different parts of the study area network to adapt the behaviour of traffic to match the primary function of the streets through which it is passing. This is found to be particularly important where traffic redistribution is not entirely feasible but where accident risk is not great enough to justify a site-specific measure. These measures are intended to complement the other measures of the scheme and often have their greatest value in satisfying local residents' concerns.

Redistribution and general treatment, when read together in an overall package of schemes, can expect to produce a saving of 10%. This is in addition to the savings expected from single site action.

Where feasible, accident reduction should be assessed by the likely effects of specific treatments. For example, if a treatment measure is designed to reduce right-turn accidents at a junction where these are particularly high, then the saving can be taken as the reduction to the normal rate for that junction type. Guidance on accident rates at different junction types is available from the TRL.

10.14.1   COST OF ACCIDENTS

Department of Transport valuations of the costs of personal injury accidents at June 1994 prices are:

Injury	£
fatal accidents	850 260
serious accidents	101 590
slight accidents	10 010
all PIAs	31 460
damage-only accidents	990
average per PIA allowing for damage-only accidents	48 930

The accident benefit of a measure is the number of accidents saved multiplied by the appropriate cost per accident.

The value of an accident remedial scheme is usually represented by its first year rate of return, which should normally be at least 15% for consideration for funding:

$$\text{first year rate of return} = \frac{\text{accident benefit for one year}}{\text{cost of implementation}}$$

10.15   Road safety plans

Virtually all local highway authorities produce annual road safety plans and guidance on their production is provided by the Local Authority Association.[8]

These plans set out the authorities' strategy for road safety and plans for safety measures. These will cover road safety engineering measures proposed but will also cover other issues, such as:

- accident data summary
- road user education
- publicity
- enforcement
- safety targets
- coordination
- encouragement of safety awareness
- monitoring.

10.16   Road safety audits

Safety audits[9,10] are primarily intended to ensure that new road schemes, improvements to highways and traffic management measures are designed and implemented to operate as safely as possible. Road safety audits are mandatory for all trunk road, i.e. DoT schemes, and are good practice for measures on all other roads.

The audits identify potential safety hazards typically under different grades of severity, such as 'problem' or 'warning'. The auditor or audit team report to the client project manager who will, when necessary, then instruct the scheme design team to respond with alternative designs.

Audits are undertaken at each of three stages for DoT schemes:

- completion of preliminary design before publication of draft orders;
- completion of detailed design before invitation to tender;
- completion of construction.

The scope of the audit becomes progressively more detailed between stages one, two and three. Items covered include:

- departures from design standards
- alignment
- junction layout
- provision for pedestrians and cyclists
- signs and lighting
- buildability
- road markings
- safety fences.

10.17 Monitoring performance of remedial measures

10.17.1 CONTROL SITES

Any monitoring of the performance of remedial measures is likely to involve the use of control sites to remove the effect of changes in numbers of accidents in the area due to other events, such as area accident trends over time or changes in traffic flow.

Control sites should be similar to the site being monitored. For example, control sites could be all other signalised junctions in the same town as the signalised junction site subject to remedial measures.

10.17.2 EVALUATING EFFECTIVENESS

Some of the criteria that can be used for evaluating effectiveness of remedial measures are:

- accident numbers
- traffic speeds
- traffic flows
- public perceptions.

10.17.3 STATISTICAL AND NUMERICAL TESTS

Statistical or numerical tests of effectiveness in reducing accident numbers include:

- chi-squared test
- standard error
- K test.

The chi-squared test has already been described (page 123). Standard error methods identify the size of the effectiveness.

The K test is a simple numerical test using data from control sites:

$$K = \frac{a/b}{A/B}$$

where

b = before-accidents at site
a = after-accidents at site
B = before-accidents at control sites
A = after-accidents at control sites.

If K is less than 1 then the site has had a decrease in accidents relative to the control sites.

References

1. Department of Transport (annual) *Road Accidents Great Britain – The Casualty Report*, HMSO, London.
2. Department of Transport (annual) *Road Accident Statistics in English Regions*, HMSO, London.
3. Institute of Highways and Transportation (1990) *Guidelines for Urban Safety Management*, IHT, London.

4.   Department of Transport (annual) *Transport Statistics Great Britain*, HMSO, London.
5.   Institute of Highways and Transportation (1990) *Highway Safety Guidelines – Accident Reduction and Prevention*, IHT, London.
6.   Royal Society for Prevention of Accidents (1992) *Road Safety Engineering Manual*, RoSPA, London.
7.   European Transport Safety Council (1995) *Reducing Traffic Injuries Resulting From Excess and Inappropriate Speed*, EC, Brussels.
8.   Local Authority Association (1989) *Road Safety Code of Good Practice*, LAA.
9.   Department of Transport (1994), HA 42/94, and *Road Safety Audits*, HD 19/94, HMSO, London.
10. Institute of Highways and Transportation (1996) *Guidelines on Road Safety Audit*, IHT, London.

11
Traffic Calming

11.1   Objectives

Traffic calming has two main objectives: the reduction in numbers of personal injury accidents, and improvement in the local environment for people living, working or visiting the area.

Traditional traffic management uses physical measures and legislation to coerce and mould driver behaviour so as to coax higher capacities out of the highway network, with improved levels of safety. Traffic calming now uses an expanded repertoire of measures and techniques to change the driver's perception of an area. Many streets portray the impression that they are vehicular traffic routes that have some other uses of lesser importance, such as shopping streets or for residential access. Traffic calming can alter the balance and impress upon the driver that the street is primarily for shopping or residential use and that vehicular traffic is of secondary importance.

Regardless of the prime cause of accidents, it has long been recognised that there is a direct correlation between accident severity and vehicle speed. Excessive speed for the prevailing road conditions can in itself be the prime cause of some accidents. Speeding traffic can cause severance effects between two parts of a community due to the difficulties experienced when pedestrians attempt to cross the road. High speed vehicles produce high noise levels and consequently degrade the environment. The TRL has estimated that a 5% reduction in accidents can be expected for every 1 km/h reduction in vehicle speeds.

11.2   Background

Almost from the dawn of the motor age, transport planners and policy makers have assigned a hierarchy to the road network with interurban trunk roads, primary distributors, district distributors, local distributors and access roads. In urban areas, increasing vehicular traffic levels and congestion has eroded the differences between the road types. Longer distance traffic has diverted to rat-runs through local areas and traffic flows have grown on secondary roads to levels formerly associated with primary routes.

Highway engineers have always tried to design roads to appropriate standards for their position in the hierarchy and road safety was always uppermost amongst their objectives. In the new towns this approach can be seen clearly. In residential areas, narrow roads with tightly curving alignments lead to wider carriageways with larger radii, until the national motorway or trunk road network is reached. Most of the basic network of roads in the older cities, towns and villages in Britain was established long before motorised transport. In the Victorian era there was a massive expansion in the sizes of settlements and wide, straight grid-like street patterns were constructed.

Diversion of traffic to parallel routes adjacent to congested routes was successfully tackled by traffic engineers in the 1970s and 1980s, using a wide range of techniques including road closures, as described in earlier chapters.

In Europe, particularly Holland, Germany and Denmark, where problems were similar to the UK, a different approach was adopted. Many of the UK techniques were used. The basic approach was a redesign of the roads in built-up areas generally but with an emphasis on residential areas. The redesign of streets and areas in Europe has not only reduced speeds and accidents by making the driver feel that he is intruding into an essentially pedestrian environment, but has softened and enhanced the aesthetics of the treated area.

In Britain, traffic calming has, with a few exceptions, continued as the imposition of traffic engineering measures to reduce speeds and accidents. The concept of redesign has not often been achieved, partly because of the legislative framework in the UK and lack of financial resources, but also due to some erstwhile highway and traffic engineers who have been slow to adopt radical new ideas.

Most traffic calming schemes in the UK have involved a form of speed-reducing road hump. Speed humps are, undoubtedly, cheap and effective in reducing vehicle speeds but it is debatable whether they have a calming effect on drivers. The demand for traffic calming schemes continues unabated, as does the search for inexpensive measures that are acceptable to drivers and effective in reducing accidents and speeds.

11.3   Site selection and ranking

Limited financial resources means that there will be competition for schemes within a local authority area. There is a danger that councillors and officers will be pressured by those who shout loudest, to implement schemes in those areas first. Prioritising of schemes, based upon objective criteria, will not only help to relieve these pressures but should result in resources being applied where most benefit can be achieved.

There are two approaches to producing a ranking for traffic calming, the single site or road and the area-wide approach. In urban areas it is likely that treatment of a single road will have a knock-on effect upon parallel routes and other adjacent streets. In small rural towns and villages where there are no nearby parallel routes, the single site can be examined and assessed in isolation.

For assessment purposes, the single site can be used successfully provided there is a clear understanding of the diversionary effect and the area affected. Preliminary cost estimates which might be used in the ranking list would need to cover the treatment of the diversionary network. If a single site is chosen for implementation it would trigger treatment of the wider network if this is necessary.

The area-wide approach would apply to an identifiable area or cell within the road network. A significant problem with the area-wide approach is that local accident or environmental hot spots can be so diluted by a safe, tranquil hinterland that they are ignored. To identify hot spots a more traditional AIP approach might be needed to highlight sites for treatment. The single site approach is less likely to miss isolated sites.

Any ranking system must have a common basis for assessment. Single site assessment can be based upon various rates per kilometre of road. This may also be a sound basis for area assessments and will tend to minimise the effects of the size and density of the road network in the area.

The prime objectives for traffic calming schemes vary from each local authority area to another. However, it is probable that limited financial resources will dictate that accident reduction is the most important, but not exclusive, objective. Table 11.1 shows some priority factors that can be used to determine a priority ranking for single site traffic calming schemes. It is possible that two parallel routes, that might be mutually affected by calming measures in the other, would appear on the same list. The route with the highest priority factor would trigger measures on both routes (and the intervening/hinterland areas).

Table 11.1 Priority factors for traffic calming schemes (adapted from that used by the London Borough of Ealing)

Criterion	Range	Priority factor
A Vehicle speed (mph)	over 45	12
(85th percentile speed)[a]	41–45	10
	36–40	8
	31–35	6
	26–30	4
	20–25	2
	under 20	0
B Vehicle flow (vehicles/hour)	Over 1000	10
(average for peak hours)	900–1000	9
	801–900	8
	701–800	7
	601–700	6
	501–600	5
	401–500	4
	301–400	3
	201–300	2
	101–200	1
	under 100	0
C Cyclists (average per hour over four highest hours in any day)	per 10	1
D Pedestrians crossing road (pedestrians/km/highest hour over four highest hours in any day)	per 100	3
E Number of frontage residents/km	per 100	1
F Accident level	per accident	5
(personal injury accidents/km/year averaged over a 3 year period)	under 1	0
G Special features in area[b]	school entrance	6
	hospital, nursing home	3
	elderly persons home	3
	bus route	3
	local shops, post office etc.	3
	doctor surgeries	3
	community centres	3
	luncheon clubs	3
	voluntary play groups	3
	nurseries	3

[a] 85th percentile speed refers to the speed at or below which 85% of traffic is travelling; this is a commonly accepted value for speed assessment studies.
[b] 'Special features' means features which are likely to create particular vehicle, and more particularly pedestrian, movements on the road network.

The area-wide approach uses a similar method to establish a hierarchy. Both approaches could include more subjective, qualitative criteria particularly where there is less pressure on resources.

The site selection process not only provides essential information to the programming of detailed study and scheme implementation. It can be used, after a settling down period, to provide a measure of the effectiveness of the implemented scheme. A new priority level can be calculated after a year (or preferably 3 years). Hopefully, the scheme, if fully implemented, will have fallen from the priority list and will only require vigilance from the traffic engineer to ensure that the implemented measures do not degrade through unsympathetic or poor maintenance.

11.4   Consultation

The questions here are when? where? who? and how to consult? There is little point trying to impose schemes upon people. It is probable that the ranking procedure will have identified areas with problems that are well known to occupants. As usual there is more than one way to address this problem:

- ask local people, by means of a questionnaire, their views on various aspects of traffic in their neighbourhood and, possibly, canvass their solutions;
- present a set of affordable potential options to the local people and ask their views on the options, traffic problems in general and the alternative solutions.

The first method ensures that local people feel involved, but can receive criticism for not explaining what could be done. The second method can receive the criticism that local people are being presented with a *fait accompli* and are not being allowed to participate sufficiently.

Fig. 11.1 Public exhibition at Peckham.

When a number of affordable options have been designed it is essential to obtain a consensus view from local people. Supporting information in the form of accident plots, speed surveys, traffic flows, photographs of trouble spots, scheme options and diagrams and photographs of features similar to those proposed should be provided (Figure 11.1). Local people are often interested in the possible after effects, costs and any example sites within reasonable travelling distance from their homes. It should be made crystal clear to consultees that it is an exercise in public participation and not just consultation, and that proposals can and will be adjusted where necessary or even dropped altogether. A carefully constructed questionnaire will help in the identification of a preferred option. Consultations must include the emergency services and public transport operators.

11.5   Traffic calming techniques

Traffic calming techniques can be broken down into four categories:

- legislation and enforcement
- surface treatment and signing
- vertical deflection
- horizontal deflection.

11.5.1   LEGISLATION AND ENFORCEMENT

These methods consist of restrictions on movement and parking including speed restriction, banned turns and one-way streets. In every case the effectiveness of each measure will depend upon the levels of enforcement that are achievable. Traffic orders made under the Road Traffic Regulation Act by the local highway authority are needed to enable enforcement to be carried out.

Enforcement of speed restrictions is aided by speed cameras but, as drivers become more familiar with them, effects are likely to deteriorate to their immediate vicinity. Camera support masts and loop detector installations cost between £5000 and £7000 per site plus the cost of the camera (£15 000–£20 000) and processing of film and prosecutions. It is common for one camera to be moved between up to eight prepared camera sites as most police forces maintain a smaller number of cameras than sites. Speeding motorists will, increasingly, become aware that the probability of being caught is small.

11.5.2   SURFACE TREATMENT AND SIGNING

Surface treatment can consist of simple coloured or textured lengths of carriageway, such as those used for gateway treatments (see page 142), through to whole blockwork paved streets. Surface treatments can be very effective but generally require reinforcement by more positive measures.

Signing is an essential – to give authority to traffic orders. It has been used with surface treatments to produce village/town gateways, which increases drivers awareness when entering speed limited areas. Signing alone cannot produce all of the calming effects required in an area and its use is governed by the *Traffic Signs Regulations and General Directions 1994*.[1] The TSRGD is comprehensive but does not allow for innovative, one-off designs to be used.

11.5.3   VERTICAL DEFLECTION

Most vertical deflections consist of some form of speed reducing road humps.[2] Road humps were first introduced experimentally in the 1980s and produced immediate speed reductions (Figures 11.2–11.4). The *Highways (Road Humps) Regulations 1996* allowed their use throughout the

Fig. 11.2 Narrowed carriageway and rumble strip, East Acton.

Fig. 11.3 Humped zebra crossing, Ealing.

Fig. 11.4 Miniroundabout and speed humps, Shoreham.

country at suitable sites and with fairly strict siting criteria.[3] The humps were round or flat topped and a series of humps had to be preceded by a speed reducing feature (such as a give-way line, a sharp bend or roundabout). Many authorities side stepped the speed hump regulations, which were considered too restrictive, and introduced speed tables and platforms where the carriageway is lifted up to footway level, usually in a contrasting colour or textured material such as concrete blocks. The 1990 Regulations brought virtually all vertical deflections under its control.

Most bus operators were unhappy with speed humps but accepted long speed tables with ramp gradients shallower than 1:15 (6.7%) or speed cushions.[4,5] Speed cushions are similar to speed humps but are narrower so that wide vehicles can traverse them with little interference. They are usually constructed in pairs but can be used singly or three abreast (Figures 11.5 and 11.6).

Other vertical deflections are: thermoplastic road humps[6] which can be between 900 mm and 1500 mm in length and up to 50 mm high and round topped; and rumble devices[7,8] which can consist of transverse strips of thermoplastic up to 13 mm high, blockwork or coarse chippings set in epoxy resin (Figure 11.7).

11.5.4 HORIZONTAL DEFLECTIONS

Chicanes can effectively reduce vehicle speeds, but designs that allow the passage of large vehicles often do not slow light vehicles sufficiently. Careful positioning of traffic islands on the approaches to chicanes can improve their performance but should be used with great care when there is heavy parking pressure. Inconsiderately parked cars can obstruct the passage of buses and other large vehicles. Traffic islands can provide a refuge for pedestrians where parking is restricted and sight lines are unobstructed. Carefully designed parking schemes can be used to

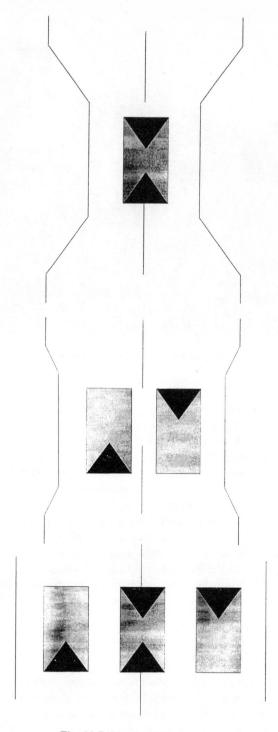

Fig. 11.5 Speed cushion layouts.

Fig. 11.6 Narrowing and speed cushions, Camden.

produce horizontal deflection of the carriageway using road markings. Adequate enforcement is essential if the beneficial effects are not to be lost.

Priority give-way designs, where one lane on a two-way carriageway is eliminated and vehicles entering an area are forced to yield to exiting vehicles is very effective. Priority give-ways have been used for village gateways and for speed-reducing features at the start of traffic calming schemes (Figures 11.8 and 11.9).

11.5.5   GATEWAYS

Carriageway narrowing combined with suitable signing, street furniture, gateposts, fencing, surface treatment or speed tables can be used to change drivers' perceptions when entering an area (Figure 11.10).[9]

11.5.6   20 MPH ZONES

The lowest national speed limit in urban areas is 30 mph but this might be higher than is appropriate in certain areas. The Department of Transport now allows 20 mph zones[10] to be implemented, generally in residential areas, where measures have been installed to both discourage through traffic and to control the 85th percentile speeds of vehicles entering and using the area to 20 mph. Zone signs must be erected at every entrance to the zone and they must be supported by a traffic order (Figure 11.11).[11]

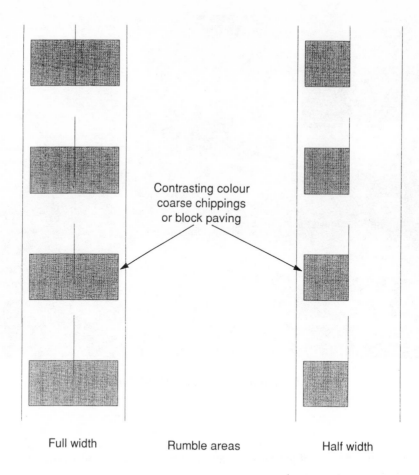

Contrasting colour
coarse chippings
or block paving

Full width   Rumble areas   Half width

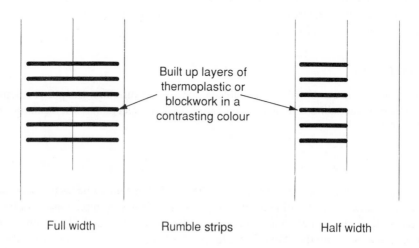

Built up layers of
thermoplastic or
blockwork in a
contrasting colour

Full width   Rumble strips   Half width

Fig. 11.7 Rumble devices.

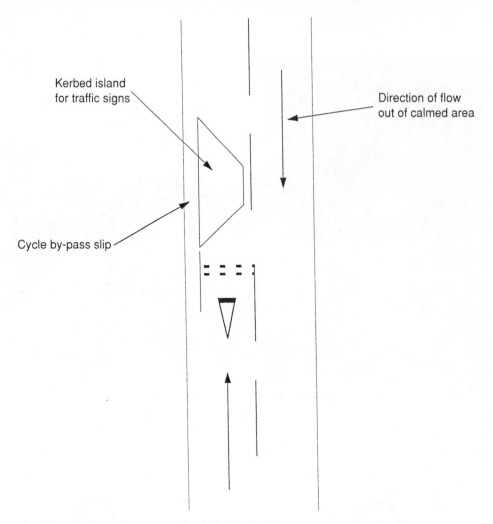

Kerbed island
for traffic signs

Direction of flow
out of calmed area

Cycle by-pass slip

Fig. 11.8 Priority give-way.

11.6   Achievable speed reductions

Vertical defections can, depending on their spacing and severity, be expected to reduce mean speeds
to 15–25 mph. Locally, at the ramp, speeds will be around 15–20 mph. Gentle chicanes and central refuges will reduce mean speeds to below 30 mph and severe chicanes to around 20 mph.

The VISP study[12,13] found that the introduction of a gateway, without any speed reducing features within the village, reduced 85th percentile speeds by between 0 and 7 mph at the gateway.
Generally, however, speed reductions were not maintained through the village if no other speed
reducing measures were introduced.

11.7   Estimate of accident reductions and benefits

The potential for accident reduction through urban safety management schemes as described in
the IHT Guidelines[14] is made up of three parts.

Fig. 11.9 Narrowing on a residential road, Shoreham.

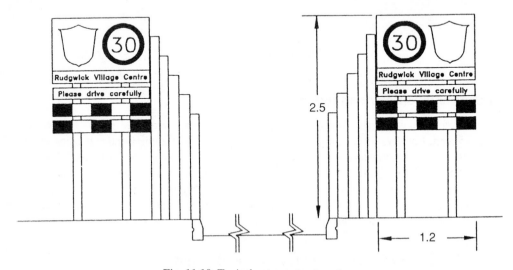

Fig. 11.10 Typical gateway treatment.

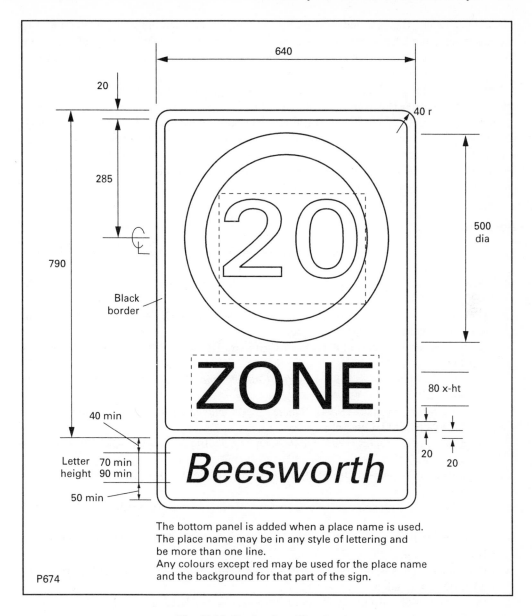

Fig. 11.11 Signing for a 20 mph zone.

The first part is a reduction in accidents achieved by traditional site-specific schemes as part of a whole area-wide package. Research has indicated that these small scale local safety measures can reduce personal injury accidents by between 15% and 80%, with an overall average of 33%.

The second part is an accident reduction achieved by measures designed to redistribute traffic within an area by actively discouraging the use of certain routes, which may not have particular accident problems but will benefit from a reduction in vehicle flow.

The third part is a reduction achieved by general treatment of different parts of the study area network to adapt the behaviour of traffic to match the primary function of the streets through which it is passing. This is found to be particularly important where traffic redistribution is not entirely feasible but where accident risk is not great enough to justify a site-specific measure. These measures are intended to complement the other measures of the scheme and often have their greatest value in satisfying local residents' concerns.

Parts two and three, when read together in an overall package of schemes, can expect to produce a saving of 10%. This is in addition to the savings expected from Part 1.

From these guidelines and the historical accident data, an estimate of the expected accident reductions can be made. If accident savings are translated into monetary values an estimate of economic benefits can be equated with the schemes' capital cost, to estimate a first year rate of return (FYRR) on the investment. This method ignores environmental benefits, which are difficult to quantify in monetary terms, and maintenance costs.

References

1. UK Government (1994) *Traffic Signs Regulations and General Directions* (Statutory Instrument 1994 No.1519), HMSO, London.
2. Department of Transport (1990) *Speed Control Road Humps*, Traffic Advisory Leaflet 2/90, HMSO, London.
3. Department of Transport (1996) *Highways (Road Humps) Regulations 1996*, Traffic Advisory Leaflet 7/96, HMSO, London.
4. Department of Transport (1994) *Speed Cushions*, Traffic Advisory Leaflet 4/94, HMSO, London.
5. London Transport Buses (1996) *Traffic Calming and Buses*, London Transport Buses, London.
6. Department of Transport (1994) *Thumps Thermoplastic Road Humps*, Traffic Advisory Leaflet 7/94, HMSO, London.
7. Department of Transport (1993) *Rumble Devices*, Traffic Advisory Leaflet 11/93, HMSO, London.
8. Department of Transport (1993) *Overrun Areas*, Traffic Advisory Leaflet 12/93, HMSO, London.
9. Department of Transport (1994) *Gateways*, Traffic Advisory Leaflet 13/93, HMSO, London.
10. Department of Transport (1991) *20 mph Speed Limit Zones*, Traffic Advisory Leaflet 7/91, HMSO, London.
11. Department of Transport (1993) *20 mph Speed Limit Zone Signs*, Traffic Advisory Leaflet 2/93, HMSO, London.
12. Department of Transport (1994) *Village Speed Control Working Group – Final Report*, HMSO, London.
13. Department of Transport (1994) *VISP A Summary*, Traffic Advisory Leaflet 1/94, HMSO, London.
14. Institution of Highways and Transportation (1990) *Guidelines for Urban Safety Management*, IHT, London.

12
Public Transport Priority

12.1   Design objectives

Public transport priority has to be seen in the context of an overall urban transport strategy with objectives which include not only improved bus (or tram) operation and restraint of car-borne commuting but also an enhanced environment for residents, workers and visitors. Measures proposed must serve all these objectives and yet also be demonstrably cost-effective and enforceable.

Typical design objectives for public transport priority measures include:

- to improve the conditions and reliability of bus operations through the introduction of appropriate bus priority measures;
- to alter the traffic balance in favour of buses at those locations where this can be properly justified;
- to improve conditions for bus passengers at stops and interchanges;
- to improve road safety generally and, in particular, for pedestrians, cyclists and people with disabilities;
- to review, where appropriate, hours of operation of waiting and loading restrictions;
- to establish and implement the coordinated and coherent application of waiting, parking and loading enforcement regimes on bus route corridors;
- to improve conditions for all road users and frontagers on bus route corridors.

Achieving these objectives often involves compromises between improving bus operation and the needs of local businesses and residents for reasonable access and of pedestrians and cyclists for safe and convenient movement.

Bus priority measures should be seen as part of the tool kit that will enable the realisation of the transport strategy. The impact of these measures on bus operation can be powerful, yet that impact should not be exaggerated. On their own, bus priority measures are unlikely to cause the major shift in travel from car to bus that is often needed to improve the urban environment. Combined with other measures, however, bus priority can contribute to a strategy of improving the urban environment and road safety and minimising the need for car travel. Typical other measures include:

- a restrictive city centre parking policy for commuters;
- improved bus services including park and ride;
- improved bus information for passengers;

- more road space provided for pedestrians and cyclists;
- traffic calming measures in residential cells;
- compatible policies for controlling new development in line with planning policy guidance as set out by the DETR.

12.2 Reliable track

The key to improved bus operation is the provision of a reliable track or right-of-way. In urban areas in the UK this track is normally shared with all other traffic and, as a result, buses experience the same congestion. This can be avoided by providing a physically segregated track, such as in busways or bus-only streets, or by reserving sections of otherwise shared track for use by

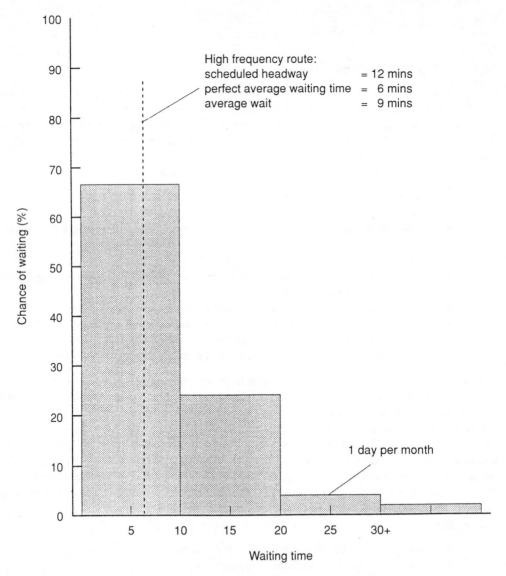

Fig. 12.1 Economic appraisal – example of bus regularity.

buses. Issues such as whether guided busways should be provided are secondary to the provision of this track.

If the track is so designed that buses can avoid congestion, then bus timings and timetables can be guaranteed with more certainty and passengers provided with a service that can be relied upon. It is the regularity and reliability of the service that is even more crucial to passengers than over-all time savings. Passengers want to know when the bus will arrive and how long it will take to reach its destination. In many congested urban areas, these two items are extremely variable.

Regularity is the variability in arrivals of buses at stops; that is the variability in bus headway. This directly affects the time passengers spend waiting at stops. Reliability is the variability in bus journey time for a given average journey; this directly affects the amount of travel time a passenger must allow to be confident of reaching the destination by a required arrival time.

Figure 12.1 shows an example, provided by a major UK bus operator, of the regularity in actual bus arrival times on a bus route compared with the timetabled arrival times. This shows a typical bus route with a frequency of five buses per hour, i.e. a scheduled headway of 12 minutes. The theoretical perfect average waiting time for a bus should be half the headway, i.e. 6 minutes. In practice, because of poor regularity, the average waiting time was recorded as 9 minutes. When considering journeys on this bus route, regular daily users would face a wait in excess of 20 minutes at least on one day per month.

12.3 Bus priority measures

Typical bus priority measures fall into four main categories:

- bus lanes and busways
- traffic and parking management measures
- traffic signal control
- bus stop improvements.

These are considered separately, but in practice the design for a bus route corridor will draw on measures from all these categories.

12.4 Bus lanes and busways

12.4.1 WITH-FLOW BUS LANES

With-flow bus lanes (Figures 12.2 and 12.3) are relatively commonplace. They enable buses to avoid queues on congested sections of road by providing a lane marked and signed clearly and implemented under a traffic regulation order prohibiting use by general traffic. Normally cyclists and taxis are allowed with buses in the lane. Sometimes commuter coaches are allowed and very occasionally, in special circumstances, goods vehicles will also be allowed.

Guidance on road markings and signing can be found in *Keeping Buses Moving*[1] and *Traffic Signs Regulations and General Directions*.[2]

Road sections are typically congested for only short periods each day and there is often a need for use of the kerbside for access to properties. For this reason, kerbside bus lanes are often restricted to operate for only a few hours each day, such as 7–10 am. Bus lanes to assist right turns would normally operate 24 hours.

The set back, length and set forward of the bus lane are important design considerations; these are illustrated in Figure 12.2. The set back defines the end of the lane and the distance back from the stop line at the next junction. For a roundabout the set back can be short, perhaps three vehicle lengths, without affecting capacity. For a signalised junction, the set back length in metres is

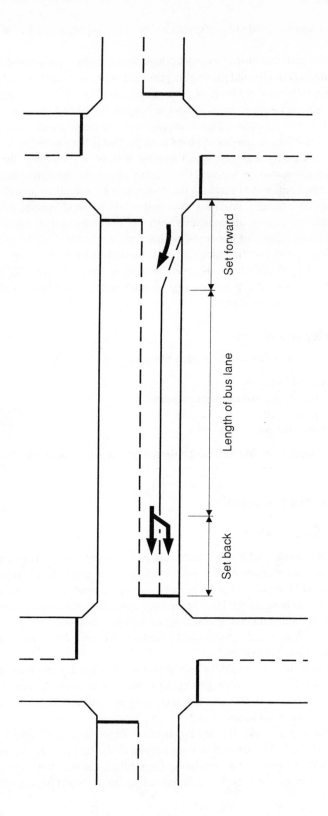

Fig. 12.2 With-flow bus lane.

Fig. 12.3 With-flow bus lane.

normally at least twice the green time for the approach in seconds to ensure that the approach capacity is not reduced. When capacity is not an issue or when buses share a lane at the stop line with left-turn traffic, then the set back can be shorter.

The length of the bus lane should be sufficient to ensure that buses avoid being trapped in a queue. If the queue with a bus lane stretches back to a previous signalised junction, then the start of the bus lane should be set forward from that junction a sufficient distance in metres, typically three times green time in seconds for a two-lane exit, to ensure that exit capacity is not reduced.

The width of the bus lane should be at least 3.0 m and a minimum of 4.0 m if cyclists are allowed to use it. Where buses need to overtake one another, as at a bus stop, then the width should be increased to 5.5 m.

Bus lanes can only be provided where the road is sufficiently wide. In a typical urban situation the road should be at least 11.0 m wide so that a goods vehicle or bus travelling in the opposite direction to the bus lane can safely overtake a stationary bus or loading vehicle.

12.4.2 *CONTRAFLOW BUS LANES*

Contraflow bus lanes (Figures 12.4 and 12.5) enable buses to avoid circuitous routes, e.g. in a gyratory system, by permitting two-way movement for buses only over a road section. The main disadvantage of a continuous contraflow bus lane is that it prevents kerbside access by

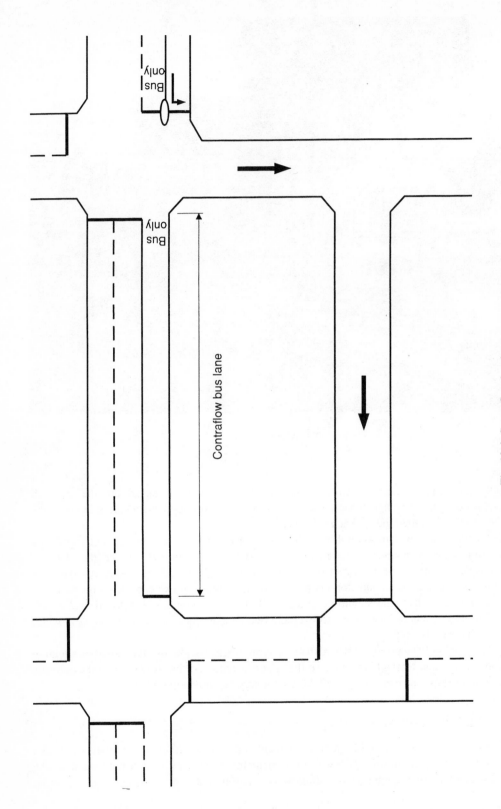

Fig. 12.4 Contraflow bus lane.

Fig. 12.5 Contraflow bus lane at Stratford.

vehicles, such as goods vehicles, that are not permitted to use it. A common method of overcoming this disadvantage, particularly when the contraflow bus lane would be long, is to use a
point closure to all vehicles except buses and then allow other vehicles access from side roads
along its length.

12.4.3   BUSWAYS AND BUS-ONLY STREETS

Busways and bus-only streets provide a dedicated track for use by buses. Busways are either separate roadways or are part of a roadway but segregated from it, normally by kerbs. One-way
busways are typically 6.0 m wide to allow for overtaking broken-down vehicles, and two-way
busways are typically 7.3 m wide.

Guided busways are an attractive option when space is limited, because the width can be
reduced as buses are guided either mechanically by horizontally mounted wheels running along
kerbs or electronically.

Bus-only streets are typically found in town centres; often there will be a speed limit of
10 or 20 mph to ensure that buses travel at speeds compatible with pedestrians. Normally
goods vehicles will be permitted access at some period during the day to load/unload at
properties.

12.5   Traffic and parking management measures

12.5.1   BUS GATES

'Bus gate' is a generic term describing all forms of control which allow buses free movement but prevent movement by other vehicles. They can literally be gates or rising bollards which open/close or lower/raise as a bus approaches and then leaves the gate. Bus gates can also be provided by traffic management measures as now described.

12.5.2   TRAFFIC MANAGEMENT MEASURES

These are typically no-entry and banned turn controls which allow buses (and possibly also cyclists) to make a movement prohibited to other vehicles.

More general measures may assist bus movement along a corridor. These measures can be side road closures, banned turns into or out of side roads, one-way streets and yellow-box markings to prevent junctions becoming blocked.

12.5.3   PARKING MANAGEMENT MEASURES

Kerbside parking is a major cause of delays to buses. Parked vehicles make it difficult for buses to approach and leave bus stops and make it necessary for buses to change lanes between stops. It is often impractical to ban kerbside parking in urban areas over the whole length of a road at all times of day, because of the lack of rear access to commercial properties requiring deliveries and the lack of off-street parking for residential properties.

Where finance and space permit, one solution is for the highway authority to enable provision of a rear access route to commercial properties and to provide dropped kerbs and allow residents to gain access over the footway to parking within the curtilage of their properties. When this solution is impractical, then it will be necessary to allow waiting and loading at certain times of day and at certain locations. Waiting and loading should be prohibited during bus operational hours at bus stops and on the approach approximately 50 m prior to the stop line at junctions that are signal controlled. At other locations waiting and loading may be permitted where essential and during hours when traffic is not congested.

In narrow shopping streets it will often be necessary to prohibit waiting and loading on both sides of the street. One solution is to provide marked bays on the roadway, staggered on alternate sides of the street. Where space permits, consideration should be given to providing bays in footway areas.

12.6   Traffic signal control

12.6.1   SIGNAL REPHASING FOR PASSIVE BUS PRIORITY

In passive priority the known volume of buses is used to alter fixed-time plan settings or SCOOT parameters. Both BUS TRANSYT and SCOOT can value buses more highly than other traffic and so give a small measure of priority to buses at intersections where other approaches have no bus flow. At many signalised intersections, all approaches are bus routes and so this method is then not effective. In addition, fixed-time coordination has to assume that buses spend a constant time at bus stops.

12.6.2   SELECTIVE VEHICLE DETECTION – ACTIVE BUS PRIORITY

In active selective vehicle detection (SVD) the presence of a bus approaching a signalled junction is detected normally by placing a detector some 60–100 m prior to the stop line. The presence of the bus then either extends the existing phase if it is currently green for the approaching bus or brings forward a green phase. The bus must be detected after any variable delay, such as at a bus stop, and this may require a trade-off with relocation of bus stops away from the optimum position for passengers. SCOOT version 3.1 includes logic to include active bus priority within a SCOOT network.

SVD can provide reductions in delay to buses at isolated signals of around 10 seconds per bus and in SCOOT a reduction of around 5 seconds per bus.

12.6.3   OVERLAP PHASES

A general concern when introducing with-flow bus lanes is that right-turn traffic would queue back beyond the set back to the bus lane. Traffic queuing adjacent to the bus lane would then be blocked and would not be able to flow into the inner kerbside lane and as a result approach capacity would be considerably reduced. One way of overcoming this difficulty is to make use of spare capacity in the direction opposite to peak flow and to provide an overlap phase for right-turn traffic in the peak-flow direction.

The overlap phase for right-turn traffic can either be provided by a late release of the opposing flow (early start of the right-turn), or an early cutoff of the opposing flow as in Figure 12.6. Research has shown that accident levels and queue lengths can be lower if late release is employed. One issue that needs consideration with late-release phases is whether to use green-arrow aspects on signal heads, as in Manchester and the North West and how to cancel them.

12.6.4   QUEUE RELOCATION AND TRAFFIC METERING

In narrow or environmentally sensitive sections of a route it can be impossible or undesirable environmentally to provide bus priority measures such as with-flow bus lanes. One solution is to relocate the queues that form in these narrow sections to wider sections on the same route. Signal control at the end of the wider section can be adjusted to meter traffic into the narrower section so that queues do not occur there. Instead queues are relocated prior to the signals over the wider road section where a bus lane can be installed. This metering of traffic is made easier in SCOOT version 2.4, which can respond to queues detected remotely from the section of road controlled by the immediate signals.

12.6.5   PRESIGNALS AND BUS ADVANCE AREAS

With-flow bus lanes are typically set back from a signalled stop lane so that approach capacity is not reduced. One way of providing further advantage to buses, particularly those wishing to make right turns, is to provide a second set of signals, the 'presignals' prior to the primary signals. The bus lane is taken to the presignals stop line.

When the presignals are red to general traffic, buses may be unsignalled or get a green phase and can move to the primary signals stop line using the area between the two sets of signals, the 'bus advance area' (Figure 12.7). The design of the bus advance area requires a carriageway wide enough for lanes to provide capacity for general traffic, a bus lane and a splitter island between the bus lane and general traffic lane(s) for mounting a signal post and head with aspects for the general traffic.

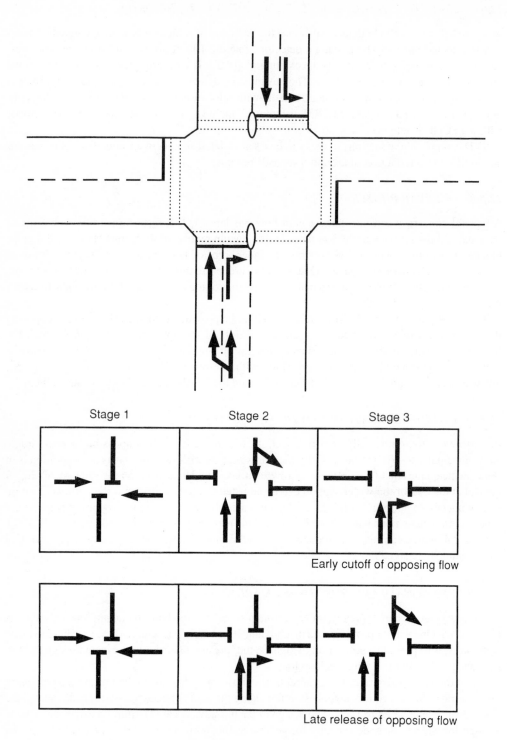

Stage 1 Stage 2 Stage 3

Early cutoff of opposing flow

Late release of opposing flow

Fig. 12.6 Overlap phase.

Fig. 12.7 Presignals – bus advance area at Shepherds Bush.

12.7   Bus stop improvements

12.7.1   BUS STOP CLEARWAYS

One of the major causes of delay in urban areas for buses and general traffic is inconsiderate parking near bus stops. Buses have difficulty in gaining access to bus stops and subsequently rejoining the traffic stream. Passengers have to resort to walking between parked cars to board a bus. Wherever possible, bus stop clearways should be introduced; these prevent waiting and loading at bus stop, typically over a 25 m length of kerb either at all times or during the working day (Figures 12.8 and 12.9).

12.7.2   BUS BOARDERS

Sometimes the imposition of waiting and loading restrictions is inappropriate because this would prevent the reasonable use of lengths of road kerb by local residents and businesses. One solution is to deter parking local to the bus stop and provide easier boarding for passengers by constructing bus boarders which consist of local extensions of the footway into the carriageway of about 1.0–2.0 m.

Fig. 12.8 Low-floor bus – double deck.

12.7.3   BUS STOP RELOCATION

The positioning of bus stops is often dictated by practicalities such as avoiding banks and post offices, where space for special deliveries is required, and avoiding creating poor sight lines for side road traffic. Bus stop locations must not unduly block general traffic, and for this reason may need to be provided with half-width (1.5 m) bus bays which give more space to general traffic while still allowing easy access and egress for buses to and from the main carriageway. If SVD at signals is required, then bus stops need to be ideally at least 70 m prior to the stop line to maximise the benefits or, alternatively, beyond the exit. Bus stops also need to be positioned to provide easy interchange between routes.

Fig. 12.9 Low-floor bus – single deck.

12.7.4 BUS SHELTERS AND TIMETABLE INFORMATION

Bus shelters are an obvious improvement at bus stops, providing protection from the weather. Timetable information should also be provided at bus stops in common static form listing the schedule of times of buses arriving at the stop and of reaching destinations, or as real-time information.

12.7.5 KASSEL KERBS

Kassel kerbs are high-sided curved kerbs which allow bus drivers to align their vehicles precisely at bus stops. The great advantage is that boarding and alighting passengers have a near-level entry and exit to the vehicle, without the need to step onto the road.

12.8 Impacts on other users

Bus priority measures should be designed to enhance the needs of other users, particularly cyclists, pedestrians, people with disabilities and frontagers.

12.8.1 CYCLISTS

Cyclists can benefit from bus priority measures both in terms of ease of travel and of safety; for example, cyclists are allowed to use with-flow bus lanes. This provides a congestion-free route and removes the risks associated with cyclists negotiating parked vehicles which would not be allowed in bus lanes. Ideally the width of bus lanes are increased at least from 3.0 m to 4.0 m to accommodate cyclists. Where this is not feasible then it will be necessary to accept that buses will need to move across the bus lane line marking so as to overtake cyclists safely. Cyclists are also, in general, permitted to take advantage of other bus priority measures such as contraflow bus lanes, busways, bus gates and presignals.

12.8.2   PEDESTRIANS

As well as improvements at bus stops, bus priority measures can provide opportunities for improving facilities for pedestrians crossing the road. When a with-flow bus lane is proposed in a road narrower than 11.0 m it may not be possible to reposition existing pedestrian refuges and so these may need to be replaced typically by zebra or pelican crossings.

The opportunity can be taken at new signalised junctions or by altered existing junctions to provide pedestrian crossing phases, normally activated by manual demand. When capacity does not permit this then with-traffic crossings with central protecting islands can be provided.

12.8.3   PEOPLE WITH DISABILITIES

Many bus operators are introducing low-floor buses on their services. This will assist people with disabilities in boarding and alighting. The design of bus priority schemes allows for this introduction by taking into account the length and swept path of low-floor buses.

12.8.4   FRONTAGERS' NEEDS

Providing priority for buses often requires changes to existing waiting and loading restrictions or the imposition of new restrictions. This must be balanced against the legitimate needs of frontagers, reflecting the different requirements of residential, commercial, industrial and retail uses. These needs are normally achieved by providing reasonable time periods when waiting and loading is permitted on the roadway of bus routes. Where this is not possible, then loading spaces can be marked on the roadway, in bays constructed adjacent to the roadway or in spaces on side-streets. Alternatively, where footways are sufficiently wide, then footway parking may be appropriate.

12.9   Process of designing and evaluating bus priority measures

Several types of survey are commonly used to assist in the design and evaluation of bus priority measures:

- bus and other vehicle occupancies
- bus journey times
- bus headways
- junction queue lengths and delays
- classified traffic counts
- parking occupancy and duration.

The design process involves consultation with bus operators, local residents, businesses and statutory bodies to identify appropriate and acceptable bus priority measures.

Junction designs and evaluations of time savings are based on the use of simulation programs such as ARCADY, OSCADY, LINSIG and TRANSYT. The assessment approach set out by the Department of Transport[1] is commonly used for the evaluation of scheme benefits. This evaluation identifies benefits such as:

- time cost savings for bus passengers and crew;
- bus operating cost savings.

Other benefits often assessed include:

- improved bus service regularity and reliability, due to less variation in travel time as the variable delays due to congestion are avoided;
- generated bus patronage due to reduced travel time and improved accessibility for bus passengers;
- time savings, enabling the bus operator to keep the same frequency of service using fewer vehicles or to improve the frequency of service with the same number of vehicles;
- decongestion where a shift to bus usage results in less car traffic and so reduces delays for all traffic.

References

1. DoT (1991) *Keeping Buses Moving*, HMSO, London.
2. DoT (1994) *Traffic Signs Regulations and General Directions*, HMSO, London.

13

The Transport Engineer and the Development Process

13.1 Planning context

13.1.1 INTRODUCTION

Most developments require planning consent from the local planning authority, i.e. the district or unitary authority, and the context for obtaining consent is described in legislation such as the Town and Country Planning Act 1990, in government policy guidance and in local authority plans. There are a few types of development that are controlled differently, such as mineral extraction which is determined at a county level, military works and major infrastructure such as strategic roads and airports.

13.1.2 GOVERNMENT GUIDANCE

Government policy guidance is contained in planning policy guidance (PPG) and regional policy guidance (RPG) notes published by the Department of Transport. There are currently some 22 PPG notes giving guidance on specific aspects of planning and some 11 RPG notes giving guidance by region:

PPG 1:   *General policy and principles*
PPG 2:   *Green Belts*
PPG 3:   *Housing*
PPG 4:   *Industrial and Commercial Development and Small Firms*
PPG 5:   *Simplified Planning Zones*
PPG 6:   *Town Centres and Retail Development*
PPG 7:   *The Countryside and the Rural Economy*
PPG 8:   *Telecommunications*
PPG 9:   *Nature Conservation*
PPG 12: *Development Plans and Regional Planning Guidance*
PPG 13: *Transport*
PPG 14: *Development on Unstable Land*
PPG 15: *Planning and the Historic Environment*
PPG 16: *Archaeology and Planning*

PPG 17:   *Sport and Recreation*
PPG 18:   *Enforcing Planning Control*
PPG 19:   *Outdoor Advertisement Control*
PPG 20:   *Coastal Planning*
PPG 21:   *Tourism*
PPG 22:   *Renewable Energy*
PPG 23:   *Planning and Pollution Control*
PPG 24:   *Planning and Noise*
RPG 1:   *Strategic Guidance for Tyne and Wear*
RPG 2:   *Strategic Guidance for West Yorkshire*
RPG 3:   *Strategic Guidance for London*
RPG 4:   *Strategic Guidance for Greater Manchester*
RPG 5:   *Strategic Guidance for South Yorkshire*
RPG 6:   *Regional Planning Guidance for East Anglia*
RPG 7:   *Regional Planning Guidance for the Northern Region*
RPG 8:   *Regional Planning Guidance for the East Midlands Region*
RPG 9:   *Regional Planning Guidance for the South East*
RPG 9a: *The Thames Gateway Planning Framework*
RPG 10:   *Regional Planning Guidance for the South West*

Other sources of official planning guidance available from HMSO include:

- Minerals Planning Guidance Notes
- Derelict Land Grant Advice
- Circulars from Government Departments
- Development Control Policy Notes
- Ministerial Statements.

The PPGs apply to England and Wales. The Scottish planning system differs slightly from that in England and Wales and so separate policy guidance applies. PPGs are to be used by local authorities in preparing development plans and in deciding on planning applications.

13.1.3   DEVELOPMENT PLANS

PPG 1 sets the scene on the planning system and confirms this as 'plan-led'. The clear intention is that proposals for development are to be judged against development plans that have been prepared following a statutory process of public consultation. These development plans include:

- structure plans – essentially these set out strategic policies typically over a county council area;
- local plans – prepared by district councils;
- waste and minerals local plans – prepared for a county area or national park area;
- unitary development plans – prepared by unitary or metropolitan authorities, such as the London boroughs.

Local plans should conform to strategic guidance in the structure plans which should themselves conform to regional planning guidance. Unitary development plans, which essentially combine local and structure plans, should also conform to regional planning guidance.

Responsibility for transport issues rests with the highway authority, being the county or unitary authority, but local plans will be prepared by the planning authority, i.e. the district councils, with guidance from the highway authority.

Proposals for development normally need to be in conformity with these plans. Proposals then will often involve first gaining acceptance in a local or a unitary development plan and, second, gaining planning consent.

Transport engineers are typically asked to assist in both these processes and sometimes also in making submissions to the structure plan consultation process if the development is of strategic importance.

13.1.4   PPG 6 AND PPG 13

The transport engineer will need to refer to the planning and regional policy guidance notes and other document sources. Two PPGs are in common use by engineers – PPG 6 and PPG 13.

PPG 6 advises local authorities on how to promote development in town centres, to select sites for development for retail, employment and leisure, and to assess retail development proposals. Some of the aims are to:

- locate major generators of travel in existing centres;
- strengthen existing centres;
- maintain and improve choice for people to walk, cycle or catch public transport;
- ensure an appropriate supply of parking for shopping and leisure trips.

PPG 13 advises local authorities on how to integrate land-use policies and transport programmes in line with the government's sustainable development strategy. The aim is to:

- reduce growth in the length and number of motorised journeys;
- encourage alternative means of travel that have less environmental impact;
- reduce reliance on the private car.

13.2   Role of the transport engineer

13.2.1   SCOPE OF ADVICE

The transport engineer will assist in the development process by advising on transport issues associated with the development, such as:

- appropriate location and type of development;
- access by all modes of transport;
- impact on the transport network;
- environmental impact;
- transport proposals;
- parking.

The engineer may be working for the planning or highway authority responsible for preparing development plans and determining planning applications. They may be working for a developer in making representations on development plans or assisting in planning applications and in implementing the development. Alternatively, they may be working for a third party in opposing development proposals in draft development plans and planning applications.

13.2.2   APPROPRIATE LOCATION AND TYPE OF DEVELOPMENT

Although the choice of location and type of development is primarily based on planning and financial issues, the transport engineer can advise on transport issues involved in sustainable development. For example, the choice of location can result in marked differences in accessibility by alternate modes of transport and in the amount of motorised journeys. The type of development on a given site can similarly affect the amount of travel; for example, a mixed land-use development or development of housing near an existing shopping or employment centre can often be shown to be more sustainable in transport terms.

In assisting in this choice, the transport engineer will be directly assessing the ability of developments to achieve the aims of PPG 13. Accessibility profiles can be of assistance in determining locational policies designed to reduce the need for travel by car.

An example of an accessibility profile, in the form of weighted accessibility indices, is shown in Table 13.1. These indices show average times to reach employment zones in Basingstoke by public transport from a set of future housing developments, both those proposed in the draft local plan and others suggested by developers. The average times have been weighted by the forecast numbers of employees in each zone.

Incorporating accessibility profiles within best practice could encourage developers to improve the profile of their site by contributing to public transport improvements. In the case of the Taylors Farm development in Basingstoke, the developer would contribute to a new rail station at Chineham on the Reading–Basingstoke line.

Another approach is to make use of a geographic information system (GIS) such as that developed for the London Borough of Croydon and shown in Figure 13.1. This will provide information on the accessibility of public transport to residents and also the relative accessibility from residential areas to attractors, such as employment areas and shopping centres. Census information is extracted using the Ed-LINE software package and road network information using the Ordnance Survey's OSCAR road centre-line software product.

The location of all bus stops has been added from London Transport BODS (Bus Passenger Origin–Destination Survey) to allow the precise calculation of walk distances. Network travel

Table 13.1 Public transport weighted journey times from housing areas to employment areas (in minutes)

	Employment areas[a]							
Housing development are as	Daneshill (8.028)	Town centre (6.150)	Viables (3.150)	Hampshire International Park (8.992)	Houndmills (8.550)	Commercial Centre (16.000)	Weighted average	Rank
1 Taylors Farm	28.0	20.0	37.0	10.0	31.0	25.0	24	1
17 North of Popley	32.0	32.0	51.0	28.0	22.0	38.0	33	3=
19 Huish Lane	35.0	38.0	57.0	63.0	53.0	48.0	49	8
20 East of Riverdene	32.0	17.0	34.0	42.0	32.0	17.0	27	2
21 Beggarwood Lane	56.0	41.0	37.0	66.0	60.0	51.0	54	11=
22 Kempshott Lane	52.0	37.0	37.0	62.0	60.0	47.0	51	10
23 Old Kempshott Lane	48.0	34.0	53.0	60.0	43.0	44.0	48	7
24 Park Prewett	37.0	30.0	49.0	37.0	25.0	40.0	36	6
30 Wimpey/Taywood Homes, Bramley	49.0	23.0	40.0	44.0	34.0	28.0	35	5
31 North Popley Fields	31.0	31.0	50.0	27.0	31.0	37.0	33	3=
32 Hodd's Farm	36.0	39.0	58.0	64.0	54.0	49.0	50	9
33 Saunder's/Hounsome	56.0	41.0	37.0	66.0	60.0	51.0	54	11=

[a] Numbers in parentheses are the weightings per 1000 employees.

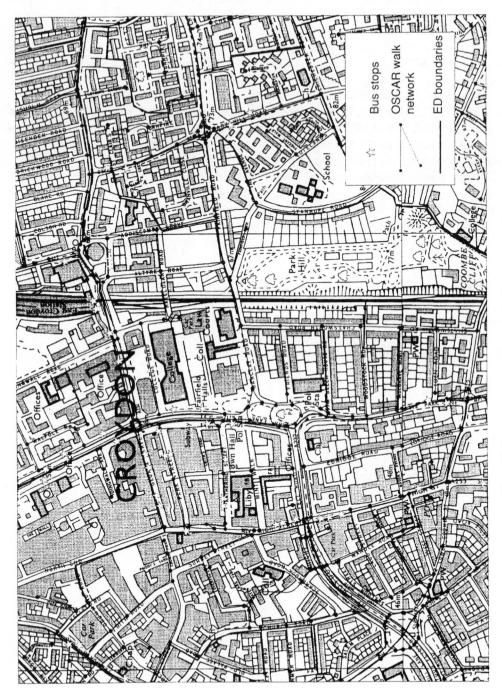

Fig. 13.1 Presentation by a geographic information system.

data can be imported from a traffic model such as that provided by the TRIPS or SATURN software referenced in Chapter 4.

In addition to providing accessibility information, the GIS can be used to produce origin–destination trip matrices, route travel times, isochrones and contours.

13.2.3 ACCESS BY ALL MODES OF TRANSPORT

In small developments, the only transport issues may be those of appropriate access to or from the highway network and off-highway parking. Requirements for these are normally set out by the local authority in published documents. Another useful compendium of information is provided by the Institution of Highways and Transportation.[1]

A new access must be located to be at least a minimum distance from other accesses. This distance depends on the role of the highway providing the access in the highway hierarchy – access road, local distributor, district distributor or primary road. The location and layout of the access must be such that it provides adequate visibility commensurate with safe stopping distances. Figure 13.2 shows the visibility requirements for a two-way straight road and for a driver eye level of 1.05 m.

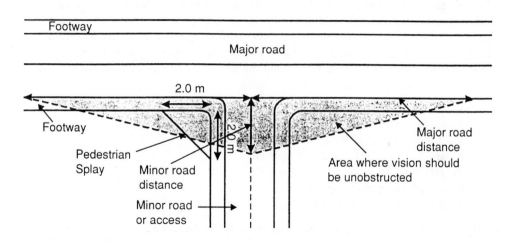

Table A: used when speed measurements are available

Major road speed (kph)*	120	100	85	70	60	50	40	30
Major road distance (m)	295	215	160	120	90	70	45	33

* 85th percentile speed, i.e. fastest speed excluding the fastest 15% of vehicles.

Table B: used when speed measurements are not available

Speed limit (kph)	70	60	50	40	30	20
Major road distance (m)	295	215	160	120	90*	45*

* Includes an allowance for motorists travelling at 10 kph above the speed limit.

Fig. 13.2 Visibility requirements.

The minor-road distance is taken as 4.5 m but can be reduced to a minimum of 2.0 m for single dwellings or a small cul-de-sac. Table A is used for calculating the major road distance when speed measurements on the major road are available; otherwise Table B is used, based on the speed limit that applies.

To protect pedestrians, there is an additional requirement for lightly used accesses as shown with a 2.0 m visibility splay at a reduced eye level of 0.6m for children.

In larger developments the form of access will also depend on capacity, environmental and road safety issues. Accesses may well be provided separately for different modes of transport.

13.3   Traffic/transport impact assessment (TIA)

13.3.1   THRESHOLDS FOR A TIA

Larger developments require an impact assessment to be provided by a transport engineer. There are no statutory rules as to when a development is sufficiently large that it requires this TIA, but the IHT document *Guidelines for Traffic Impact Assessment*[2] suggests thresholds based on either traffic flows or scale of development.

The suggested traffic flow threshold is when traffic to or from the development exceeds 10% of the two-way traffic flow on the adjoining highway. If this highway is or will become congested in the assessment period (typically 10 years) then the IHT suggest a lower threshold of 5%. An even lower threshold is likely to be required for developments affecting trunk roads.

An alternative approach is to apply thresholds either by size of development for the different use classes shown in Table 13.2 or by traffic generation or parking criteria:

- residential development (C3) in excess of 200 units;
- business (B1 and B2) gross floor area (GFA) in excess of 5000 m^2;
- warehousing (B8) GFA in excess of 10 000 m^2;
- retail (A1) GFA in excess of 1000 m^2;
- 100 trips in/out combined in the peak hour;
- 100 on-site parking spaces;
- landfill sites or quarries generating HGVs.

The IHT guidelines unfortunately concentrate on highway aspects and treat other modes as of secondary importance. The guidelines provide very useful information on traffic impact analysis but do not comprehensively cover aspects that are needed for a transport appraisal or impact assessment.

13.3.2   SCOPING STUDY

A key element of a TIA is the scoping study. The transport engineer should always aim to reach agreement between developer and highway authority on this prior to the assessment work commencing. Without this agreement the TIA is likely to prove unsuitable or will require time-consuming modification. The scoping study will include:

- size and nature of development;
- size of study area and its network;
- other transport and development proposals that need to be taken into consideration;
- transport data and models available and need for further surveys;
- assumptions on network traffic growth;

- planning standards and policies that apply;
- assumed year of opening of development and its phasing;
- years for assessment.

The size and nature of the development is likely to be well defined by the time a TIA is prepared as the definition will be included in the planning application which the TIA will support. In a feasibility study for a development, the size and nature of the development may not be defined, but it may still be appropriate to have the scope of the study agreed between developer and highway authority.

The size of the study area and its network will depend on the expected traffic generated by the development and current congestion on the network. It may be necessary to include congested junctions remote from the network. The size may also depend on other transport and development proposals known to the highway authority.

It is important to establish the availability of existing traffic data and information on other development proposals, for example TIAs prepared for other developers. Existing transport models can provide useful information and may be required when assessing future transport movements. The need for further surveys for highway traffic movements, public transport patronage and walk/cycle flows should be established. Surveys of journey times and queue lengths may also be required.

Table 13.2 Guide to use classes order

Use classes order 1987	Description
A1	Shops, retail warehouses, hairdressers, undertakers, travel and ticket agencies, post offices, etc.
	Pet shops, cats-meat shops, tripe shops, sandwich bars
	Showrooms, domestic hire shops, funeral directors
A2	Banks, building societies, estate and employment agencies
	Professional and financial services, betting offices
A3	Restaurants, pubs, snack bars, cafés, wine bars, shops for sale of hot food
	Shops selling and/or displaying motor vehicles
	Launderettes, dry cleaners, taxi businesses, amusement centres
B1	(a) Offices, not within A2
	(b) Research and development, studios, laboratories, high tech
	(c) Light industry
B2	General industry
B3–7	Special industrial groups
B8	Wholesale warehouses, repositories
C1	Hotels, boarding and guest houses, hostels
C2	Residential schools and colleges
	Hospitals and convalescent/nursing homes
C3	Dwellings, small businesses at home, communal housing of elderly and handicapped
D1	Places of worship, church halls
	Clinics, health centres, creches, day nurseries, consulting rooms
	Museums, public halls, libraries, art galleries, exhibition halls
	Nonresidential education and training centres
D2	Cinemas, music and concert halls
	Dance, sports halls, baths, skating rinks, gymnasiums
	Other indoor and outdoor sports and leisure uses, bingo halls, casinos
	Theatres

Ideally, historical traffic data will provide information on traffic growth. Alternatively, network traffic growth, that is growth that will occur without the development, can be assessed using the Department of Transport's National Trip End model. Allowance should be made for the lower growth in traffic at peak periods, mainly due to capacity constraint, compared with the growth in daily traffic.

Clarification of the applicable planning standards and policies will be needed, because these can be in the process of being changed as local authorities prepare new development plans.

Phasing of a major development may be an important consideration as it can determine the years for assessment of traffic and also trigger requirements for transport improvements. Assessments are normally undertaken for the year of opening of the development (or years of opening of major phases) and for a future year, usually 10 or 15 years after opening.

13.4   Elements of a TIA

The TIA should cover the following:
- existing conditions
- proposed development
- trip generation, distribution, modal split and assignment
- transport impact
- transport proposals
- environmental impact.

13.4.1   EXISTING CONDITIONS

Site visits will identify the nature and condition of the current transport network. The information to be collected will include road geometric data and junction layout, bus and rail services, pedestrian and cyclist facilities. The visits will also assist in assessing current transport problems, such as highway congestion and lack of adequate public transport service.

Existing traffic data will be identified and analysed, and the results of the further surveys initiated as part of the TIA will be described.

Accident data may be analysed and any local accident problems identified.

Other developments with planning consent (but not yet implemented) and planned transport improvements should be listed.

13.4.2   PROPOSED DEVELOPMENT

The proposed development will be described including:

- current or former use of site and any relocation proposals;
- planning policies applying to this site;
- any previous planning applications;
- site area and boundary;
- scale and size of land use proposed;
- hours of operation;
- development phasing;
- site plan for development.

13.4.3   TRIP GENERATION, DISTRIBUTION, MODAL SPLIT AND ASSIGNMENT

Techniques for estimating travel demand are described in Chapter 4. There is now a considerable amount of information on the amount of highway traffic generated by new developments. The TRICS database[3] is one well-established national database which holds survey information from hundreds of sites covering many different land uses. These land uses include retail superstores, retail parks, warehouses, offices, business parks, industry, leisure and housing, often with sub-categories for each usage.

Unfortunately, information on other modes of travel (i.e. public transport, walk and cycle) generated by developments is very sparse. It is often necessary to resort to more general sources of travel information such as the *National Travel Survey*.[4]

When using information from the existing databases or other sources then the approach to estimating trip generation is one of comparison. The development proposal is compared with similar development for which traffic data are available on the basis of floor area, employment or other appropriate variable.

If suitable comparative information is not available, then trip generation can be assessed either from first principles, by making broad assumptions on travel behaviour, or by surveying travel movements at a similar site. This latter method is particularly important in countries outside the UK where comparative data may be unavailable. Typical trip generation rates are shown in Table 13.3 for peak hour and total daily traffic.

Table 13.3 Typical trip generation rates

Land use	Peak arrivals (am)	Peak departures (am)	Peak arrivals (pm)	Peak departures (pm)	Total daily arrivals	Total daily departures
Offices, trips per 100 m²	1.5	0.1	0.1	1.1	4.8	4.8
Business parks, trips per 100 m²	1.2	0.1	0.2	0.9	4.0	4.0
Warehousing, trips per 100 m²	0.3	0.1	0.1	0.3	2.1	2.1
Residential, trips per household	0.2	0.5	0.5	0.2	3.9	3.9
Industrial, trips per 100 m²	0.7	0.2	0.2	0.6	4.2	4.2
Hotels, trips per bedroom	0.2	0.2	0.2	0.2	3.2	3.2
Retail parks, trips per 100 m²	0.5	0.2	0.8	1.0	12.2	12.2
Supermarkets, trips per 100 m²	2.4	0.7	6.2	6.4	68.0	68.0

The estimation of distribution and modal split can be achieved using a full transport model, as described in Chapter 4, or by a simpler process – either based on existing traffic movement data or by using isochrones or gravity distribution assumptions.

Existing traffic movement data can be obtained from a survey of movements at an existing similar nearby land use. Alternatively, census journey to work tabulations may provide a guide.

With a gravity model (page 41), population and employment data and information on journey time or distance is input to the model formula to estimate the distribution of trips to or from a generating development over the surrounding area. The form of the deterrence function is sometimes simplified to either (travel time)² or (travel distance)² if information on travel cost is not available, so that:

$$T_{ij} = P_i\, G_j\, \alpha\, /(\text{travel time})^2$$

where

T_{ij} = the trips between origin zone i and destination zone j

P_i = the population of zone i

G_j = the trips generated by the proposed development in zone j
α = has a balancing value to ensure that

$$\sum_i T_{ij} = G_j$$

An alternative, even simpler approach is to establish isochrones of equal travel time from the development. By assuming a trip-length distribution (i.e. the number of trips within each time band) from a similar site, the number of trips can be distributed into each time band using the isochrones.

When considering retail developments it is often desirable to make use of a shopping model which includes other retail stores and to assess demand for all competing stores. The output from such a model will include the forecast distribution of journeys to the development.

The forecast of trips generated by a retail development will also include an analysis of pass-by and diverted trips, being those journeys that already exist on the road network and will be broken to include a visit to the development.

If a sophisticated model is not available then modal split will have to be estimated from existing data surveys, such as observations from similar sites or from census data.

Trip assignment models are described on page 42. In TIA work manual assignment is often more appropriate than using a computer model as the number of zones is sufficiently small for a manual approach to be practical.

13.4.4   TRANSPORT IMPACT

The transport impact conventionally concentrates on highway link and junction analysis but should also include road safety and bus and rail demand.

Link flows after development should be compared with values in DoT TA46/57, *Traffic Flow Ranges for the Assessment of New Rural Roads*. This will provide an assessment of the operational performance of road links. Junctions are conventionally assessed using traffic simulation programs such as ARCADY, PICADY, OSCADY, LINSIG and TRANSYT. Merging, diverging and weaving performance on and between slip roads may also need checking. All these analyses are discussed in Chapter 5.

The highway analysis will normally be summarised as a set of reserve capacities and queue lengths at critical locations for the assessment years appropriate to the development.

The likely impact of the development traffic on road safety will be assessed, based on the examination of historical accident data.

The impact of demand on existing bus and rail services may also need to be assessed to determine any shortcomings in existing provision or any overloading that is likely to occur with the development.

The requirement for parking and any impact on neighbouring streets will be assessed.

13.4.5   TRANSPORT PROPOSALS

Transport proposals will be put forward to mitigate any negative transport and environmental impacts likely to be caused by the development. These proposals will include improvements to:

- accesses
- highway junctions and links
- pedestrian and cyclist routes and provision
- public transport services, priority and facilities
- facilities for the disabled

- parking
- road safety
- internal road layout.

The proposals will themselves be assessed to measure their performance in mitigating impacts or in providing improvements affecting existing residents and businesses.

13.4.6   ENVIRONMENTAL IMPACT

This is covered in more detail in the next section but will include:

- noise
- vibration
- community effects, e.g. severance
- air quality.

13.5   Environmental assessment

13.5.1   FORMAL REQUIREMENTS

The need for a formal environmental assessment (EA) is set out by the DoE.[5]

An EA is required if the particular development proposal is likely to have significant effects on the environment. Some guidance is given on thresholds for different types of development needing an EA. Those most likely to affect traffic engineers are:

- manufacturing industry – sites greater than 20 ha;
- industrial estates – sites greater than 20 ha or with more than 1000 dwellings within 200 m of the site boundary;
- urban developments – sites greater than 5 ha or with more than 700 dwellings within 200 m of the site boundary or providing more than 10 000 m² gross floorspace of shops, offices or other commercial use;
- urban roads – where more than 1500 dwellings lie within 100 m of the centre line of the road;
- other projects – sites greater than 100 ha.

The EA involves collecting information on the likely environmental effects of the development on human beings and on flora, fauna, soil, water, air, climate, landscape, material assets and cultural heritage. The assessment includes indirect effects such as those caused by traffic.

The planning authority will advise whether an EA is required. If it is required then an environmental statement containing the information collected during the EA is to be provided with the planning application.

13.5.2   ENVIRONMENTAL EFFECTS CAUSED BY TRAFFIC

Most developments will not require a formal EA. Even so, there may well be a need for the assessment of environmental effects caused by traffic. The Institute of Environmental Assessment has published its own guidelines[6] on how this is to be assessed and prepared the following thresholds to identify road links and areas needing assessment:

- road links where traffic flows or numbers of HGVs increase by more than 30% as a result of development traffic;

- sensitive areas, such as hospitals, conservation areas or areas of high pedestrian flow, where traffic flows or numbers of HGVs increase by more than 10% as a result of development traffic.

The 30% guideline is approximately equivalent to an increase in noise level at residential frontages of 1dB(A) which is about the smallest change in level discernible to people.

The guidelines recommend that assessments be undertaken in the year of opening of the development and over time periods when the environmental impact is greatest; these periods may not coincide with periods of maximum traffic flow.

The items to be measured include:

- noise assessed near buildings;
- vibration in buildings;
- visual obstruction and intrusion;
- community effects such as severance caused by difficulty in crossing by road or by the physical barrier created by the road;
- delays to drivers and pedestrians;
- pedestrian amenity, fear and intimidation;
- accidents and safety;
- hazardous loads;
- air pollution;
- dust and dirt, e.g. quarrying sites;
- ecological effects;
- heritage and conservation areas.

Fig. 13.3 Impact of construction and streetworks on traffic.

Procedures for measurement are included in the DoT manual on environmental assessment, included as Volume 11 in the *Design Manual for Roads and Bridges*.[7] Further technical background information is available in other documents.[8]

Mitigating effects to minimise environmental impact will include capacity improvements, pedestrian crossings, restriction on movements of HGVs (both hours and routeings) noise barriers and traffic calming.

References

1. Institute of Highways and Transportation (1997) *Transport in the Urban Environment*, IHT, London.
2. Institute of Highways and Transportation (1992) *Guidelines for Traffic Impact Assessment*, IHT, London.
3. JMP (undated) *TRICS – A Trip Generation Database for Development Control*, JMP, London.
4. Department of Transport (undated) *National Travel Survey*, HMSO, London.
5. Department of the Environment (1989) *Environmental Assessment – A Guide to the Procedures*, HMSO, London.
6. Institute of Environmental Assessment (1993) *Guidelines for the Environmental Assessment of Road Traffic*, IEA, London.
7. Department of Transport (1993) *Design Manual for Roads and Bridges volume 11: Environmental Assessment*, HMSO, London.
8. Morris, P and Therivel, R (1995) *Methods of Environmental Impact Assessment*, University College, London.

14

Designs for Sustainable Development

14.1   Sustainable development

14.1.1   INTRODUCTION

A widely agreed definition of sustainable development is 'development that meets the needs of the present generation without compromising the ability of future generations to meet their own needs'.[1]

The Government Panel on Sustainable Development[2] considers that the DoT's projected growth of road traffic over the next decades is unsustainable. The panel supports the views of the Royal Commission on Environmental Pollution[3] on the need for clear objectives and quantified targets with an increased emphasis on the role of public transport.

These are hardly new views. In the early 1960s, Buchanan[4] noted that cars are inefficient in use of space, both on the road and at the destination, and that it would be physically impossible to accommodate everyone in the city centre if they were all to travel by car.

It was also clear, by observing the impact on small and medium towns and cities in the USA, what happens to the vitality of town centres if major generators of travel demand are located on the edges of towns. Yet this is precisely what has happened in the UK since the early 1980s.

The joint DoE/DoT publication *Planning Policy Guidance – Transport*[5] (PPG 13) might then be considered as a classic case of attempting to lock the gate after the horse has bolted.

14.1.2   GOALS AND POLICIES

PPG 13 provides guidance on how local authorities can adopt policies to meet the Government's commitment to a sustainable development strategy. These policies are intended to contribute to the goals of:

- improving urban quality and vitality;
- achieving a healthy rural economy and viable rural communities.

Figure 14.1 shows some of the linkages between the strategy and the adopted policies in summarised form. The implication is that it is essential that land usage and transport needs be planned together, to contribute to the sustainable development strategy.

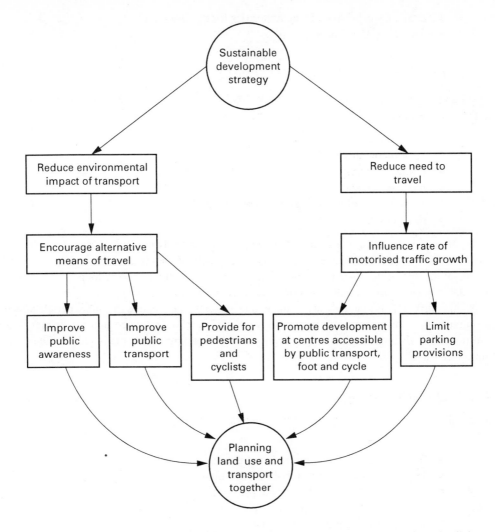

Fig. 14.1 Some of the linkages between sustainable development strategy and the adopted policies.

14.1.3   OPPORTUNITIES AND CONSTRAINTS

Planners and engineers will seek the appropriate opportunities to fulfil these policies but will be subject to a number of practical constraints.

The opportunities are:

- change land-use and transport patterns;
- manage traffic demand;
- develop transport policies and proposals through the planning process;
- contribute to public awareness of the unsustainable consequences of present actions.

The constraints are:

- public resistance to change of mode of transport;
- demand for choice of mode of transport;

- competition between local authorities for development;
- need for urban regeneration.

14.1.4 CHANGE LAND-USE AND TRANSPORT PATTERNS

The Standing Advisory Committee on Trunk Road Assessment[6] (SACTRA) observed that road improvements induce additional traffic due mainly to:

- total volume of activities
- location of activities
- timing of activities
- mode of transport used
- coordination of activities by different individuals
- route chosen.

By altering the location of activities and by reducing the utility of travel by car, it would be the aim to achieve a reduced traffic effect; that is the opposite of the effects observed by SACTRA.

One of the difficulties is that changing land-use patterns will take a very long time because of the relatively slow turnover, 30 years or more, in reuse of land.

Another difficulty is that there are so many out-of-town retail centres that more people now use them rather than town centres for their weekly convenience shopping needs. It can be argued that additional out-of-town retail centres for convenience shopping may actually reduce the need for motorised travel.

14.1.5 MANAGE DEMAND

Some measures, well-known to transport planners and engineers, for managing demand for car traffic are:

- improve heavy rail and introduce light rapid transit (LRT);
- improve bus frequency, reliability and quality of service;
- implement bus and LRT vehicle priority;
- reduce public transport fares;
- increase fuel charges for motorists;
- traffic calming and environmental traffic management;
- introduce park and ride schemes;
- implement pedestrianisation schemes and area bans on certain vehicles;
- improve facilities for pedestrians and cyclists;
- improve information on public transport services.

These measures are all part of the toolkit for managing demand, particularly in urban areas. To be effective, these measures need to be coupled with either parking restraint or congestion charging.

Parking restraint implies:

- control of both number and price of public on-street and off-street parking;
- limiting the number of spaces allowed in new developments;
- possibly controlling existing private nonresidential parking.

Congestion charging would be achieved in towns and cities by one of the forms of urban road pricing or area licensing and between towns and cities by motorway tolling.

14.1.6   DEVELOP TRANSPORT POLICIES AND PROPOSALS THROUGH THE PLANNING PROCESS

PPG 13 makes it clear that the transport policies and proposals should aim to:

- support the locational policies and objectives of the plan (local, unitary or structure);
- improve the environment;
- reduce accidents.

Local authorities drawing up their transport policies and programmes[7] are expected to demonstrate that their proposals harmonise with the guidance in PPG 13. PPG 13 also implies that there should be no protection of road corridors for schemes on which work will not commence within the plan period. This may effectively rule out the planning of new roads in urban areas to cater for the long-term traffic growth projected by the DoT.

PPG 13 highlights the desirability of avoiding loading primary roads with short-distance local traffic and of avoiding accesses from new developments directly onto primary roads. Land-use policies should then support transport aims, including safeguarding the role of primary roads.

14.1.7   CONTRIBUTE TO PUBLIC AWARENESS

Transport planners and engineers can contribute to public awareness of the issues involved through initiatives such as Hampshire County Council's HEADSTART. The public in many towns is aware, through personal experience, of the effects of congestion, which have been assessed by the CBI at £15 billion per year in the UK. The environmental effects of transport may not be so obvious. It has been suggested that the health cost of particulate emissions from vehicles is around £14 billion per year in the UK.[8]

Short trips in cars are particularly emission inefficient, partly because of the limitations of catalytic converters.[9] If new development is concentrated in existing centres, as recommended by PPG 13, then short trips should increase as a percentage of all trips. It is these short trips, of length less than 5 miles, that can be readily transferred from car to walk, cycle or bus. Public awareness of the issues could enable this transfer.

The Countryside Commission[10] has also pledged to increase public awareness of the impact of traffic growth upon the countryside as well as publishing practical guidance on rural traffic calming and demand management.

14.2   Public resistance to sustainable development policies

14.2.1   CAR TRAVEL AND CHANGE OF MODE

The privacy, practicality, quality and convenience of car travel provides a very obvious resistance to change of mode from car to public transport, walk or cycle.

Bus patronage has declined in the UK overall by about 22% between 1985 and 1994. A study of five towns[11] showed that if this lost patronage could be recovered, then buses would carry 16–44% of urban travellers using only 3–11% of the roadspace. Several improvements are necessary to reverse the decline in bus usage:

- bus priority to improve reliability and journey times;
- modern vehicles to provide a high quality travelling environment and full access for those with disabilities;

- better information about services;
- better terminal and bus stop facilities;
- improved frequency of service.

Some reasons for resistance to change to walk and cycle are:

- lack of safe, direct, well-lit routes and paths for cyclists;
- lack of secure, weather-protected cycle parking;
- lack of safe routes for pedestrians, particularly children.

The latter point is particularly important and has been a major reason for the dramatic reduction over the last 20 years in the percentage of children walking to school. So often, highway authorities only improve pedestrian facilities in response to accidents and in line with DoT guidelines rather than as a response to encouraging walking.

Countries such as Germany, Netherlands and Denmark have achieved a greater public acceptance of public transport, walk and cycle modes than the UK. The Royal Commission of Environmental Pollution[3] aims to move the UK in the same direction and has recommended the setting of targets to increase public transport use from 12% at present to 30% by 2020 and to increase cycle use of all urban journeys from 2.5% at present to 10% by 2005.

14.2.2   DEMAND FOR CHOICE

Living in a market economy we expect to have choice and freedom of movement. This is true for seeking employment opportunities, shopping or leisure activities. Pricing or other demand management controls on transport need to be seen by the public as providing benefits that outweigh the disadvantages.

14.2.3   COMPETITION BETWEEN LOCAL AUTHORITIES

Local authorities compete to maintain or improve their economies. This may lead to the offering of incentives to employers to locate in an authority's area by granting employers the numbers of car parking spaces they desire.

It can also lead to authorities setting parking charges in their town centres at a low level or providing further parking spaces to attract shoppers and visitors away from town centres in other authorities. There will inevitably be a conflict between PPG 13 and further provision of parking spaces for shoppers.

Parking policy can clearly, for these reasons, have repercussions well beyond the boundary of an individual local authority. The government has stated that in the South East[12], authorities should cooperate to develop parking policy at a subregional level.

14.2.4   NEED FOR URBAN REGENERATION

Many authorities will in practice give the need for urban regeneration, and the resultant creation of employment opportunities, a higher priority than reducing congestion or reducing pollution. Prospective employers currently often demand sufficient parking space so that all employees can drive to work. PPG 13 must not be considered to negate the opportunity for urban regeneration.

14.3   Responses by transport planners and engineers

14.3.1   GOVERNMENT POLICY

Transport planners and engineers have witnessed a major shift in government policy towards surface transport since 1995. PPG 13, although setting out the new stand on transport planning issues, is surely not the Government's last word on the subject.

Consultants have been employed to prepare a good practice guide[13] and to monitor how effectively PPG 13 is being implemented in practice.[14] More fundamentally, the Government Panel on Sustainable Development[2] recognises 'the Government's wish for a broad national debate on the subject followed by a thorough and measured response'.

14.3.2   SUSTAINABLE DEVELOPMENT MEASURES

Some of the opportunities available for enabling the sustainable development process are:

- local employment and shopping opportunities;
- walkway and cycleway networks;
- roads designed to deter short distance journeys by car;
- public transport improvements;
- creation of a car-free culture;
- telematics applications.

If employment and shopping opportunities are available close to residential areas then there is an increased likelihood of journeys being contained within the immediate area. This will encourage short distance journeys. If a network of walkways and cycleways is provided then these short distance journeys are more likely to be made on foot or by cycle.

In a new housing development much can be done to aid journeys by foot or cycle if the road layout discourages short distance journeys by car within the development area and to the local centre.

Public transport improvements and locations of developments at nodes well geared by public transport will encourage a higher usage of public transport services. If linked with demand management measures, then modal share by public transport can realistically be increased.

A car-free culture can be created by a combination of these measures with a travel awareness campaign.

Telematics applications are discussed in more detail in Chapter 15 but can minimise the need for travel and provide information for more efficient journeys.

14.3.3   PERFORMANCE INDICATORS

The performance indicators that may be used by transport engineers for measuring the sustainability of development proposals include:

- containment of journeys within a development area;
- daily car journeys per household;
- daily car kilometres per household;
- modal share;
- road congestion reference flows;
- impact on adjoining settlements;
- viability of public transport services.

14.4   Design for pedestrians

Pedestrians are a vulnerable group of road users. Designing for their needs requires a careful consideration of road safety and provision of adequate capacity and ease of use.

Two very useful publications dealing with pedestrian crossings have been published by the Government as Local Transport Notes.[15,16] They highlight the key issue of pedestrian crossings. Another important issue is that of walkway capacity.[17]

A typical walkway ultimate capacity is 56 pedestrians per minute per metre, whereas a more comfortable value of 28 pedestrians per minute per metre gives a Fruin level of service C, defined as 'speeds generally restricted and minor reverse flows would encounter some difficulty'.

A reduction factor of up to 15% is applied if flows are not balanced in the two directions. Allowance is made for obstructions and boundary effects by reducing overall width to effective width. Typically, boundary effects are taken as 0.3 m and items such as handrails or guard rails as an additional 0.3 m.

Improving walkway widths will provide a better level of service for pedestrians and also reduce the time needed to cross the road. The resultant reduction in roadway width may often be achievable without sacrificing highway capacity and with the additional benefit of reducing traffic speeds.

Pedestrianisation of streets completely removes the possibility of vehicle conflict. Where complete pedestrianisation is not feasible or desirable, then the street can be shared with other vehicles using appropriate surface treatment and speed controls. When flows are moderate and speeds kept low, then buses can successfully share the street with pedestrians without creating an atmosphere that prevents free movement by pedestrians. Other vehicles, such as delivery vehicles, can share the street at specific periods during the day.

Crossings should be located where pedestrians need them, wherever safe to do so. This may seem obvious but is not always achievable as a safe distance needs to be maintained between crossing and side roads and between crossing and the exit from a roundabout so that drivers can stop safely. Sometimes this constraint can be overcome by banning turning movements out of a side road or by making it one-way away from the junction.

A site assessment will provide the information needed to design a crossing and will typically include:

- pedestrian demand flows;
- time to cross including delay before crossing;
- vehicle flows;
- vehicle speeds;
- accident records;
- surrounding land uses;
- physical road layout including nearby junctions;
- any visibility constraints for vehicles and pedestrians, such as nearby parking;
- other road features – lighting, surfacing, bus stops, parking.

The assessment should recognise that the provision of pedestrian crossings should improve safety and also the perception of safety, so that journeys on foot can be encouraged.

An assessment framework can then be used to determine which form of crossing is appropriate. Crossing types include:

- refuge islands
- traffic calming measures
- zebras

- pelicans
- puffins
- toucans
- staggered crossings
- other signal control.

Visibility requirements for all types of crossing are given in Table 14.1.

Table 14.1 Visibility requirements for all types of crossing

85th percentile approach speed (mph)	25	30	35	40	45	50
Desirable minimum visibility (m)	50	65	80	100	125	150
Absolute minimum visibility (m)	40	50	65	80	95	115

Refuge islands are often the cheapest solutions but restrict roadway width and require waiting and loading restrictions on the adjoining kerb. The Local Transport Note suggests an absolute minimum of 1.2 m width for this island but anybody who has pushed a pram knows this to be woefully inadequate!

Traffic calming measures can both lower vehicle speeds and narrow the roadway that has to be crossed by pedestrians. Drivers are required to stop when they see pedestrians on zebra crossings and so this can allow pedestrians to establish precedence when vehicle speeds are low. The Local Transport Note recommends that zebra crossings should not be installed on roads with an 85th percentile speed of 35 mph or above as pedestrians are unlikely to feel sufficiently confident to be able to establish precedence. Another practical point is that zebras can cause undue delay to vehicles if both vehicular and pedestrian traffic is heavy and will not then be the best choice of crossing.

In these two situations a pelican crossing will be more suitable. This type of crossing normally has vehicle actuation and gives priority alternatively to vehicles and pedestrians with pedestrians being shown a green man/red man signal. The level of priority is determined by the vehicle precedence time, set between 20 and 60 seconds, which is the maximum time that vehicles continue to have priority after the pedestrian push-button is operated. The green man walking time is fixed in the controller and is set depending on the carriageway width to be crossed.

Puffin and toucan crossings have extended this logic. Puffins (pedestrian user-friendly intelligent signals) include kerbside and on-crossing detectors. Kerbside detectors provide for cancelling pedestrian demands if no pedestrian is observed waiting to cross the road; this prevents unnecessary operation of a pedestrian phase as occurs with pelicans. On-crossing detectors can extend the all-red time when a pedestrian is still on the crossing; this can occur with the elderly and mobility handicapped. Toucans provide a shared crossing for both pedestrians and cyclists.

Staggered crossings are recommended on two-way roads wider than 15 m to reduce crossing distances, but may be advantageous for widths down to 11 m. The stagger should be left-handed so that pedestrians waiting to cross on the central refuge island should always be facing the traffic. The central island should be provided with safety barriers and be at least 3 m wide, although there is a special design in use in London when this width is not available.

Other forms of signal control can provide crossing time for pedestrians. One common method is to include pelican crossings within an urban traffic control system or link them to nearby signal-controlled junctions. The pedestrian phase, if called, will then be preset by the UTC or linked-control settings.

Even more common are pedestrian crossing facilities at a signal-controlled junction or very close to it. Under these circumstances the crossing phase can be either preset or called by push-button control; pedestrians are offered a green man aspect for crossing but a blanked-out instead of a red man aspect for closures of the phase.

Most other countries in the world do not have pelican or similar crossings but rely on controlled junction crossings. In some countries, such as the USA, this is aided by a right-turn rule (equivalent to left-turn in the UK) when this turning traffic must give way to pedestrians crossing the adjacent roadway. This is a great advantage as it avoids complex phasings and use of additional refuge islands often required in UK designs.

14.5   Design for cyclists

Cycling had, until recently, become the forgotten mode of transport in the UK. There is now a widespread interest in encouraging cyclists and a number of useful documents[18-21] give guidance on designing for cyclists.

The importance of cycling as a mode of travel in the UK is very patchy. Nationally only some 2.5% of journeys are made by bicycle, but in cities where cycling is positively encouraged, such as York, Oxford and Cambridge, the percentage can be as high as 20%.

Cycling is often said to be a dangerous mode of travel and UK statistical evidence indicates this to be correct. This danger is due to a lack of awareness and consideration by many drivers and to a lack of cycle-friendly roads. This is highlighted by the Danish experience. In 1970 Denmark had the worst record for cyclist safety in Western Europe. After a positive programme of changes, adult Danes now cycle 12 times[22] the distance of British adults and their safety record, measured by distance travelled, is 10 times better.

The road safety issues surrounding cycling should be addressed by public awareness campaigns and improved design. There is a strong argument that improved health, freedom of movement and accessibility available through cycling outweighs any increase in injuries.

Traffic engineers can provide the design improvements leading to safe routes for cyclists (Figures 14.2–14.4) through measures such as:

- cycle tracks
- shared facilities with pedestrians

Fig. 14.2 Advanced stop line for cyclists, Manchester (courtesy of Department of Transport).

- reduced vehicle speeds and flows
- modified designs at roundabouts
- priority at signalised junctions
- better surfaces
- better parking.

   The starting point is an assessment of demand for cycling bearing in mind Government's policy of increasing usage. Conventional transport models can be modified to quantify the demand and to assist in identifying routes for cyclists.

Fig. 14.3 Pedestrian and cyclist crossing (courtesy of Department of Transport).

Fig. 14.4 Cycle bypass.

Where land is available it may be possible to provide cycle tracks. These should be a minimum of 2 m wide, be provided on both sides of a main road and be designed to provide priority across side roads. This latter can be achieved by bending the track away from the carriageway so that cyclists can see and be seen by turning traffic and by providing humped crossings.

Cycle tracks can often be shared with pedestrians in which case this width should desirably be at least 3 m. Shared routes should be signed. It is not normally necessary to segregate cyclists and pedestrians on shared routes except where these are steep and cyclists' speeds are likely to cause pedestrians to feel uneasy. Where segregation is provided then this can be a white line rather than a kerb so that passing movements are not restricted.

Broadly speaking, cyclists can ride safely with vehicular traffic at speeds below 20 miles per hour. Above this speed additional lane width is desirable, particularly if traffic flows are heavy. When traffic speeds are above 30 mph some form of segregation or additional lane width is preferable. For speeds above 40 mph segregation is required. Traffic calming measures, described in Chapter 11, can reduce speeds and also remove through traffic from streets in residential or sensitive areas.

A particular issue occurs at priority junctions with large mouths where drivers continue at high speed endangering cyclists. Corner radii can be reduced to minimise this danger.

Roundabouts which allow high speeds are particularly dangerous for cyclists. Solutions include redesigns to lower speeds, signalisation or provision of segregated cycle routes. Miniroundabouts are safer but flared entries encouraging high speeds should be avoided.

Advanced stop lines can help cyclists to make right turns safely at signal-controlled junctions and to minimise conflict with left-turning vehicles.

Priority on road links can be provided by enabling cyclists to share bus priority measures and make use of safe exemptions from one-way street and banned-turn controls, or by providing cyclist-only entries to streets. With-flow cycle lanes have an obvious appeal but need to be at least 1.5 m wide and to be supported by waiting and loading restrictions. Mandatory cycle lanes are marked with a solid white line and vehicles may not cross them. Advisory cycle lanes, marked with an intermittent white line, are used where vehicles need to cross the line, for example to make turns.

Two other issues that are also important for cyclists are surfaces and parking. Surfaces should be smooth (e.g. asphalt and not cobbles or paving) and well drained. Dropped kerbs should be used where cyclists move from roadway to track. Cycle parking needs to be provided close to destinations and in secure convenient areas using cycle stands, such as Sheffield stands, or lockers.

References

1. Brundtland Commission (1987) *Our Common Future*, OUP.
2. Department of the Environment (1995) *British Government Panel on Sustainable Development, First Report*, HMSO, London.
3. Royal Commission on Environmental Pollution (1994) *Eighteenth Report: Transport and the Environment*, HMSO, London.
4. Buchanan, C (1963) *Traffic in Towns: A Study of the Long-term Problems of Traffic in Urban Areas*, HMSO, London.
5. Departments of the Environment and Transport (1994) *Planning Policy Guidance – Transport* (PPG 13), HMSO, London.
6. Standing Advisory Committee on Trunk Road Assessment (1994) *Trunk Roads and the Generation of Traffic*, HMSO, London.

7. Department of Transport (1994) *Transport Policies and Programme Submissions for 1995–96*, Local Authority Circular 2/94 DoT, London.
8. Baker (1995) Sustainable Development Balancing Act Waits for Government Lead, *Surveyor*, 23 Feb.
9. Royles (1995) *Literature Review of Short Trips*, Project Report 104, TRL, Crowthorne.
10. Countryside Commission (1993) *Sustainability and the English Countryside*.
11. Confederation of Passenger Transport (1994) *The Role of the Bus in the Urban Economy*.
12. Department of the Environment (1994) *Regional Planning Guidance for the South East*, DoE, London.
13. Department of the Environment (1995) *A Guide to Better Practice*, PPG 13, HMSO, London.
14. Departments of the Environment and Transport (1995) *Implementation of PPG 13 Interim Report*, DoT/DoE, London.
15. Department of Transport (1995) *The Assessment of Pedestrian Crossings*, Local Transport Note 1/95, DoT, London.
16. Department of Transport (1995) *The Design of Pedestrian Crossings*, Local Transport Note 2/95, DoT, London.
17. Fruin, J (1971) *Pedestrian Planning and Design*, Metropolitan Association of Urban Designers and Environmental Planners, New York.
18. Institute of Highways and Transportation (1996) *Cycle Friendly Infrastructure – Guidelines for Planning & Design*, IHT, London.
19. Sustrans (1994) *Making Ways for the Bicycle*, Sustrans, London.
20. Sustrans/Arup (1996) *The National Cycle Network*, Sustrans, London.
21. Automobile Association (1993) *Cycling Motorists: How to Encourage Them*, AA, Basingstoke.
22. Mynors, P and Savell, A (1992) *Cycling on the Continent*, Travers Morgan, London.

15

Transport Telematics

15.1 Introduction

Transport telematics, also known as intelligent transport systems (ITS), is concerned with the application of electronic information and control to improve transport. Some new systems have already been implemented and the pace of implementation can be expected to quicken. With a crystal ball, we can foresee how a typical journey to work may look in 10 years time.

Before leaving home, you check your travel arrangements over the Internet. Often you choose to travel by public transport and you can identify travel times and any interruptions affecting the service. On this occasion, you choose to travel by car as you have an appointment later in the day at one of those old-fashioned business parks that are inaccessible by public transport. There are no incidents recorded on your normal route to work so you do not bother to use your computer route model to select an optimum route for you.

Once in your car, you head for the motorway and select the cruise control, lane support and collision avoidance system, allowing you to concentrate on your favourite radio service. Suddenly, this is interrupted by the radio traffic-message channel service giving you information about an incident on your route. You are not surprised when, at the next junction, the roadside variable message sign (VMS) confirms this; motorway messages really are believable now!

You feel pleased with yourself that you have precoded your in-car navigation system with the coordinates of your final destination, and soon you are obtaining instructions on your best route with information updated from the local travel control centre.

As you near your place of work, you are aware of roadside messages informing you of the next park and ride service. You choose to ignore these as you will need to make a quick getaway for your appointment. You then check that your travel card is clearly displayed inside the car; you don't want to be fined for not having a positive credit for the city's road pricing and parking service! The same card gives you clearance to your parking space; you activate your parking vision and collision control just to be sure of not scratching the MD's car next to you.

15.2 Using transport telematics

All these information and control services, and many more besides, are discussed in the UK Government's consultation document.[1] One way of categorising these services is into the following application areas:

- traffic management and control
- tolling and road pricing

- road safety and law enforcement
- public transport travel information and ticketing
- driver information and guidance
- freight and fleet management
- vehicle safety
- system integration.

All these applications are being developed with assistance from research and pilot implementation programmes in Europe, USA and Japan.

15.3 Traffic management and control

Any traffic management and control system needs information on traffic flows, speeds, queues, incidents (accidents, vehicle breakdowns, obstructions) air quality and vehicle types, lengths and weights. This information will be collected using infrared, radio, loop, radar, microwave or vision detectors. In addition, public and private organisations will provide information on planned events (roadworks, leisure events, exhibitions).

The use to which this information is put depends on the objectives set for management and control. Network management objectives set for urban areas[2] include:

- influencing traveller behaviour, in particular modal choice, route choice and the time at which journeys are made;
- reducing the impact of traffic on air quality;
- improving priority for buses and LRT vehicles;
- providing better and safer facilities for pedestrians, cyclists and other vulnerable road users;
- restraining traffic in sensitive areas;
- managing demand and congestion more efficiently.

The software systems used will include control applications such as SCOOT, SCATS, SPOT and MOTION. These are responsive systems which control a network of traffic signals to meet these objectives. Automatic vehicle location and identification will provide information for giving priority or allowing access to certain vehicles only.

Air quality monitoring can be linked to a control centre to provide information for decisions on changes to traffic control.[3]

Interurban network management systems will have similar objectives but will make greater use of access control by ramp metering and other means, and of speed control and high-occupancy vehicle lane management. Regional traffic control centres will advise motorists of incidents and alternative routes by VMS and by RDS-TMC, a signal FM radio service broadcasting localised traffic messages and advice to drivers.

Examples of integrated interurban and urban management systems are provided by Munich's COMFORT,[4] Southampton's ROMANSE project and by MATTISSE (Figure 15.1) and the Midlands Driver Information System (Figure 15.2).[5]

15.4 Tolling and road pricing

Interurban motorway tolling and urban road pricing provide another approach to meeting network management objectives while obtaining additional revenue that can be invested in transport. Singapore's electronic zone pricing, the TOLLSTAR electronic toll collection and ADEPT automatic debiting smart cards are examples of such applications.

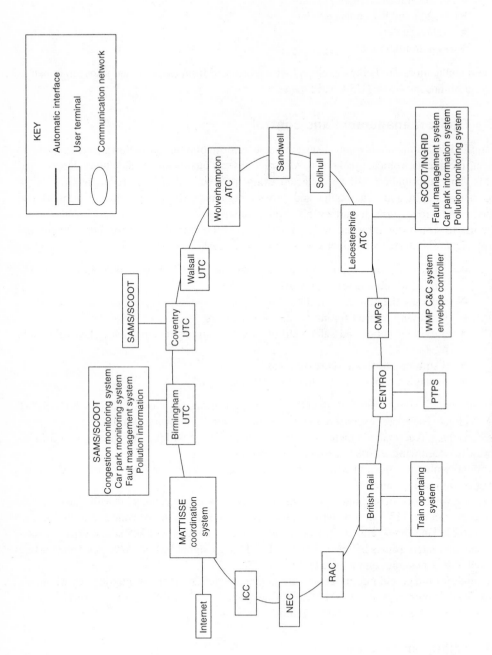

KEY

Automatic interface
User terminal
Communication network

SAMS/SCOOT
Congestion monitoring system
Car park monitoring system
Fault management system
Pollution information

SAMS/SCOOT

SCOOT/INGRID
Fault management system
Car park information system
Pollution monitoring system

WMP C&C system
envelope controller

PTPS

Train opertaing system

Internet

ICC

NEC

RAC

British Rail

CENTRO

CMPG

Leicestershire ATC

Solihull

Sandwell

Wolverhampton ATC

Walsall UTC

Coventry UTC

Birmingham UTC

MATTISSE coordination system

Fig. 15.1 MATTISSE system configuration.

Fig. 15.2 Motorway variable message sign.

These systems rely on microwave or radio communication to an in-vehicle transponder in a smart card with detection of vehicle licence plates using image processing for enforcement purposes.

15.5   Road safety and law enforcement

Excessive speed is a major factor in at least one third of all road accidents. It is not surprising then that telematics applications for road safety have concentrated on measurement of speed and detection of drivers breaking speed limits. Initial installations have involved film cameras, but video equipment with a secure telemetry connection to a control centre is likely to become more popular in the future.

Red-light-running cameras are a similar application for capturing records of drivers ignoring traffic signal red lights.

A more exciting prospect being considered in the EC MASTER project is to use on-board vehicle electronic speed limiters linked to throttle and brakes for ensuring speed limits cannot be breached. Roadside radio or microwave transmitters would send speed limit information to passing vehicles. This potentially would also readily allow the introduction of variable speed limits by time of day or according to weather conditions.

Another possible in-vehicle system based on smart cards and electronic vehicle identification would guard against vehicle theft and unlicensed vehicles.

Other enforcement systems[6,7] can detect offences in bus priority measures, such as illegal entry or parking in bus lanes.

15.6   Public transport travel information and ticketing

Travel information is needed by passengers at home or office and also during their journey. London Transport's ROUTES computer-based service offers routeing, timetable and fares information on all public transport services in London through public inquiry terminals.

Real-time travel information is provided in London by the COUNTDOWN system which is being expanded to cover 4000 bus stops. A similar system called STOPWATCH is available in Southampton as part of the ROMANSE project and is based on Peek's Bus Tracker system which can detect buses using either radio beacons or GPS (Global Positioning System) which uses satellites to identify locations.

ROMANSE also includes TRIPlanner interactive enquiry terminals with touch screens providing travel information.

Problems with tickets for through journeys can be a deterrent for travellers choosing public transport. Smartcard stored-value tickets can provide a single ticket for car parking and all legs of a journey served by different operators.

15.7   Driver information and guidance

Driver information systems include the RDS-TMC radio data system-traffic message channel, initially trialled between London and Paris in the PLEIADES project and elsewhere in Europe in similar EC-funded projects. There is also the Trafficmaster service which uses infrared monitors to identify congestion and an in-car visual map-based screen to inform drivers of congestion.

Driver guidance systems aim to take this a step further by informing drivers of their route and giving guidance on navigation. Communication between the control centre and the vehicle can be by roadside beacon or by digital cellular radio networks based on GSM (global system of mobile communications) as in SOCRATES. Commercial products include Daimler Benz's copilot dynamic route guidance system trialled in Berlin and Stuttgart and Philip's Car Systems CARiN. Similar products, such as the VICS advanced mobile information service, are commonly available in Japan.

15.8   Freight and fleet management

Applications for freight and fleet management are based on automatic vehicle location using GPS. Both vehicles and loads can be tracked for security and for optimising fleet usage.

If destinations are known in advance, then the operator can schedule routes and time deliveries more efficiently. The operator can also identify opportunities for consolidating loads or obtaining return loads.

15.9   Vehicle safety

Many vehicle manufacturers or systems suppliers (Ford, Daimler Benz, Lucas, Jaguar) are developing vehicle safety applications.

Autonomous Intelligent Cruise Control uses microwave radar sensors to ensure a safe time interval between cars. This could be used to allow cars to travel close together safely, thus improving road capacity.

Antilock braking systems can now be enhanced with electronic traction control to prevent wheel spin.

Lane support systems use a video camera and image processor to detect kerb and lane markings and then to warn the driver if he is about to stray from the lane.

Collision warning and avoidance systems detect obstacles, warn the driver and take action to avoid collision.

Driver monitoring will monitor the way in which a driver is controlling the vehicle and provide a warning if performance indicates drowsiness or loss of attention.

15.10   System integration

Many of these applications can benefit from common standards and a convergence of system architectures. This is because many of these applications use the same technologies and need to pass messages between detectors, control centres, vehicles and roadside information and control systems or between applications.

Data exchange is a key area and the DATEX standard provided by the EC covers this.

In the UK, a planned motorway architecture is provided by NMCS2 and a planned urban traffic management architecture should emerge. Private-sector systems providers will have a strong role in setting future standards and architectures.

15.11   International comparisons

Miles[8] has researched ITS developments in several countries including the UK, mainland Europe, USA and Japan.

In the UK, the DoT is promoting an open modular architecture for urban traffic management and control (UTMC). The full list of functions that this will support is shown in Figure 15.3. The Highways Agency is aiming to obtain private-sector involvement in regional traffic control centres (RTCCs). These centres will control and manage trunk roads and provide information to drivers about roadworks, congestion and incidents.

In Germany, both Munich through the COMFORT project and Stuttgart through the STORM project have developed ITS to an advanced state. The system coverage includes:

- integrated traffic and travel data centre
- public transport information and management
- park and ride information
- UTC and regional traffic control
- beacon and RDS-TMC traffic information services
- freight management.

In Turin, Italy, a complete integrated road transport environment (IRTE) has been set up to connect 10 traffic and transport control centres to a backbone data communications network. The centres have connections to a town supervisor with a common database and a set of forecasting models to improve performance.

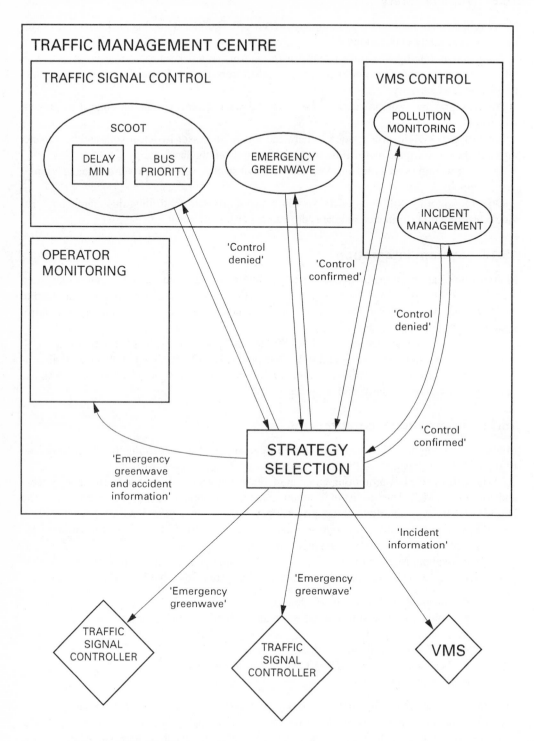

Fig. 15.3 Urban traffic management and control strategy and selection function.

The USA has defined an intelligent transportation infrastructure (ITI) and deployment has occurred in four cities: Phoenix, San Antonio, Seattle and New York. ITI covers nine core technologies:

- regional multimodal traveller information centres
- traffic signal control systems
- freeway management systems
- transit management systems
- incident management programmes
- electronic fare payment systems
- electronic toll collection systems
- highway–rail crossing protection
- emergency management services.

In Japan, the National Police Agency is developing a universal traffic management system (UTMS). The system uses an infrared detector which is used for both control and for two-way communications (Figure 15.4). The down-link to the vehicle is capable of carrying traffic information for the driver, whereas the up-link carries a vehicle identifier permitting journey time monitoring.

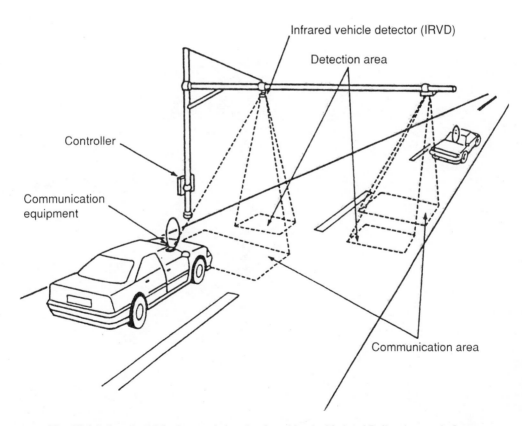

Fig. 15.4 Infrared vehicle detector being developed by the National Police Agency in Japan.

References

1. Department of Transport (1996) *A Policy for Using New Telematic Technologies for Road Transport – Consultation Document*, HMSO, London.
2. Routledge, Kemp and Radia (1996) UTMC: The Way Forward for Urban Traffic Control, *Traffic Engineering and Control*, November.
3. Hewitt, R (1996) ELGAR, *IEE Conference*, London, 1996.
4. Hoops, Csallner and Busch (1996) Systems Architecture for Munich COMFORT, *Traffic Technology International*.
5. Welsh, P and Carden, P (1997) MATTISSE and the Midlands Driver Information System, *TRAFFEX Conference*, PTRC.
6. Hewitt, R, Slinn, M and Eastman, C (1996) Using Cameras to Deter the Illegal Use of Bus Lanes in Birmingham, *IEE Colloquium on Camera Enforcement of Traffic Regulations*, London, 1996.
7. Turner, D and Monger, P (1996) The Bus Lane Enforcement Cameras Project, *IEE Colloquium on Camera Enforcement of Traffic Regulations*, London.
8. Miles, J (1997) UTC Meets ITS; what does the future hold for Urban Traffic Control Systems?, *TRAFFEX Conference*, PTRC.

16

Statutory Requirements

16.1 Introduction

Traffic engineering activity, particularly on the highway, is controlled by an enormous number of laws and regulations which, taken together, set out:

- the rights of landowners;
- the rights and responsibilities of road users;
- the rights and responsibilities of the highways authority and, by inference, the traffic engineer acting on their behalf;
- the powers of the police to enforce the law, particularly as it relates to traffic and travellers.

In the UK, legislation is formulated in a number of ways. Acts of Parliament or primary legislation provide the main source of the laws that govern roads and traffic. Often, however, Acts of Parliament only provide general and nonspecific enabling powers. These powers allow something to be done, but without specifying the manner in which that thing should be done. In these circumstances secondary legislation, in the form of a Statutory Instrument, is required to give effect to the primary legislation contained in a Parliamentary Act.

A good example of this two-tier relationship can be found in the *Road Traffic Regulation Act 1984*.[1] This Act gives a highway authority the power to erect traffic signs but makes no provision as to the form or type of signs allowed. These have been subsequently defined in a Statutory Instrument SI 1994/1519 *The Traffic Signs Regulations and General Directions 1994*.[2]

This two-tier system of legislation avoids detailed technical matters becoming the subject of an Act of Parliament. It also means that the legal requirements can be adapted more quickly in response to changing circumstances and changing technology, without the need to enact new primary legislation, provided that the underlying need remains the same. Thus, for example, the 1994 Signs Regulations replaced an earlier 1981 Statutory Instrument, reflecting the change in sign designs over the intervening years, and the introduction of brown tourist signing, but without a change to the underlying requirement for signing.

The government also issues countless circulars, advisory notes, guidelines and technical papers which formally set out guidance on the interpretation of the law and advise on the procedures to be followed in implementing new systems of traffic management and control.

The government acts to regulate and control highways and traffic on a national basis. However, it is also possible for a local authority to seek legislation which has only a local effect.

Thus, for example, it is an offence to park on the footway within the Greater London area, because of a local act (*The GLC General Powers Act 1969*)[3] and, although equivalent powers exist nationally, in the *Road Traffic Act 1974*,[4] the Government has never brought the powers into effect. Local powers can be granted by means of a local government act which applies specifically to an area, or by means of a local by-law.

16.2   Scope of legislation

The legislative process is ongoing and it is inevitable that whatever is written here will be overtaken by events very quickly. Recognising this, the following attempts to set out the main principles of legislation and to identify the key acts and regulations which set out the main body of relevant law. With an extensive body of written law, the tendency in recent years has been to only bring forward primary legislation when it is required to legislate in some new area, or to update or remove law which has become obsolete.

The main areas of legislation of relevance to the traffic engineer are:

- the Highways Acts, which mostly deal with the provision and maintenance of highways and roads;
- the Road Traffic Acts, which set out the laws controlling road traffic, often including laws relating to vehicles and drivers;
- the Road Traffic Regulation Acts, which primarily set out the powers that a highway authority has to regulate and control traffic on roads.

16.3   Highways Acts

The *Highways Act 1980*[5] provides many of the basic laws relating to the provision and maintenance of highways. The words highways, roads and streets are in common usage and are used interchangeably by most people. However, the words have different meanings in law, although many may find the definitions and differences are far from clear.

16.3.1   HIGHWAY

We might reasonably expect to find a workable definition of a highway in the Highways Acts. The converse is true. The *Highways Act 1980* is far from helpful saying that 'a highway means the whole or part of a highway other than a ferry or waterway'! A more useful definition is offered in *An Introduction to Highway Law*[6] which suggests that, based on common law, a highway is, in simple terms, a route which everyone can use whenever they wish, without let or hinderance and without charge.

The only exceptions to the legal requirement that highways should be available for use without charge are those few highways which are subject to special Parliamentary Acts, such as the Dartford Tunnel and Bridge, where the highway was created by a special Act which included the power to charge a toll for use of the facility. Historically this type of highway, which has usually been an esturial crossing, has been provided using a special Parliamentary Act. However, the *New Roads and Street Works Act 1991*[7] created a more general power which allows a highway authority to enter into an agreement with a third party, called a concession agreement, to allow the third party to build a new road and then levy a toll for its use. The concessionaire would be free to set the toll at whatever level they wished, except for an esturial crossing, where the government can set an upper limit on the toll.

16.3.2   ROAD

The legislation is similarly vague in offering a simple definition of a road. The *Road Traffic Regulation Act 1984* says that a road means 'any highway or of any other road to which the public has access'. Thus a highway is a road and a road is a road but, as far as the law is concerned, a road is only a road if the public has access over it.

This is an important difference. The public have a right to use a highway. However, on a road, which is not a highway, the public may have access but not necessarily the right to pass and repass, without let or hinderance. This definition could apply for example to a service road on a retail development, where the public are allowed access, but not as a right. Although not offering a very useful definition of what a road is, this definition has the useful effect of allowing the application of road traffic law to roads, such as service roads and private roads which are not highways.

16.3.3   STREET

The legal definition of the word street given in the *Highways Act 1980* and repeated in the *New Road and Street Works Act 1991* is:

    (a) any highway, road, lane, footway, alley or passage;
    (b) any square or court; and
    (c) any land laid out as a way, whether it is for the time being formed as a way or not.

This perhaps could be interpreted as any land where the use of the land is to provide a means to get to adjoining property. Thus streets may not be roads, or highways but could give access to property.

16.4   Road Traffic Acts

The Road Traffic Acts deal with drivers and vehicles, and specify:

- the offences connected with the driving of a vehicle;
- the qualifications required for drivers;
- the construction and use of vehicles.

The latest Road Traffic Act, the *Road Traffic Act 1991*,[8] in addition to creating a number of offences also created a new decriminalised system of parking. The law allows local highway authorities to take over responsibility for the enforcement of most stationary vehicle offences from the police. The Act also moved the offences from the ambit of the criminal law to the civil law, with the penalty dealt with as a debt, rather than as a fine.

The *Road Traffic Act 1991* also created a new administrative system for the main road network in London. The Act allowed the creation of the Priority route network with the position of Traffic Director for London to manage the network.

The Act also set out the powers which provided for the use of red-light and speed cameras to record, respectively, drivers who fail to stop at a traffic signal or who are speeding.

16.5   Road Traffic Regulation Act

The *Road Traffic Regulation Act 1984* provides the legal basis for management of the highway network. The main parts of the Act are:

Part I   Provides the general powers for the regulation of traffic and gives powers which allow most parking and traffic control functions.

Part II   Deals with special types of traffic regulation.

Part III   Deals with pedestrian crossings and playgrounds.

Part IV   Deals with provision of parking.

Part V   Provides the powers which allow a highway authority to provide traffic signs.

Part VI   This contains the powers relating to controlling the speed of traffic.

Part VII   Although there is a basic right to 'pass and repass' on the highway, the highway authority has powers, under this part of the Act, to erect obstructions, for example bollards, to limit the types of traffic that can use a highway.

Parts VIII
and IX   Deal with enforcement powers.

These powers generally apply to England, Scotland and Wales. There are separate powers in Northern Ireland.

16.6   Other legislation

The form and structure of the UK legislative system means that powers which affect the highway can appear in any Parliamentary Act and it would be impossible to authoritatively identify all the relevant statutes in a work of this sort.

There is, however, a body of law dealing with the rights of other bodies, such as the public utilities, to install and maintain their services in the highway. The traffic engineer needs to be aware of the rights and responsibilities of these organisations.

The *New Roads and Street Works Act 1991* deals with two areas of legislation. The Act provides powers for the provision of new roads, financed by the private sector. The Act allows that bodies, other than the highways authority, can provide a new road and recover the cost of that road by charging a toll for its use.

Parts III and IV of the Act are concerned with street works. Prior to the passing of this Act, the many agencies with access to the highway could, and did, carry out their activities in a fairly uncoordinated way, often with little thought as to how their actions might affect traffic movement. The rationale of the 1991 Act is to provide a basis for regulating and managing the work of highways authorities and statutory undertakers, so that, as far as is practical, activities can be coordinated so as to minimise their adverse impact on travel.

References

1.   *Road Traffic Regulation Act 1984*, HMSO, London.
2.   UK Government (1994) *Traffic Signs Regulations and General Directions 1994* (Statutory Instrument 1994 No 1519), HMSO, London.
3.   *Greater London Council General Powers Act 1969*, HMSO, London.
4.   *Road Traffic Act 1974*, HMSO, London.
5.   *Highways Act 1980*, HMSO, London.
6.   Orlik, M (1993) *An Introduction to Highway Law*, Shaw and Son.
7.   *New Roads and Street Works Act 1991*, HMSO, London.
8.   *Road Traffic Act 1991*, HMSO, London.

Index